TREE

A NEW VISION *of* THE AMERICAN FOREST

For Suzanne, Simone, and Emily

For the tree of life that allowed these pictures to bloom

TREE | A NEW VISION *of* THE

AMERICAN FOREST | JAMES BALOG

foreword by

DAVID FRIEND

STERLING PUBLISHING
New York

FOREWORD Out on a Limb *David Friend*

ALL AROUND US CREATION BLOOMS, unbidden and without witness. Forces beyond our vision—or even our imagining—are perpetually at work and at play. Our spiral Milky Way, imperceptibly swirling, supports our sun that tugs our earth that tugs our moon that tugs our ocean tides that disgorge the spiraled seashells that we hold up to our ear so as to hear, as the poets put it, the music of those spheres and the echoes of that ocean. Responding to cues not yet explained by either science or poetry, cells divide, tectonic plates shift, and arctic terns migrate, making their annual trek—some 25,000 miles—without the indulgence of AAA or GPS. Our own lives, too, are the stuff of shadow puppetry, our actions swayed by genetics and society, psyche and circumstance.

And then there are the secrets of the trees.

Though trees are the natural world's goliaths—the largest living things, the most resilient, the oldest (some literally outlasting millennia)—much of their magic plays out unseen. "Every year," according to Annie Dillard in *Pilgrim at Tinker Creek*, "a given tree creates absolutely from scratch ninety-nine percent of its living parts. Water lifting up tree trunks can climb one hundred and fifty feet an hour….A tree stands there, accumulating deadwood, mute and rigid as an obelisk, but secretly it seethes." Observed John Muir, the naturalist who helped champion Sequoia and Yosemite National Parks, "To the outer ears these trees are silent, [yet] their songs never cease. Every hidden cell is throbbing with music and life, every fiber thrilling like harp strings."

These unobserved processes, of course, take place within the recesses of every tree. Trees possess subterranean root systems that in some sense mirror the flaring of their branches. Trees track the march of the seasons, etching each year's passage in covert grooves and glyphs, called rings. And each tree can house its own ark's worth of species. (Photographer Gary Braasch, on assignment for *Life*, shot for seventeen days in the limbs of a 200-foot sloanea tree in the Costa Rican rain forest, documenting some of the eight hundred forms of animal and plant life, indigenous to the tree, all going about their cryptic business.)

And even in their dying, trees persist. Washington State photographer and educator Steve Wilson has tried to demystify decay by taking pictures of the same fallen Douglas fir over the course of forty years. "The decay of a single tree," contends Wilson, "can take hundreds of years. Once it is horizontal and on the forest floor, this community of life that had been vertical and seeking nutrients suddenly *becomes* nutrient. Its bark is gnawed by chewers. Micro-foragers savage for food. And the process of life continues."

No wonder, then, that amid all its hidden treasures (ecological, biological, teleological), an individual tree often gets lost in the proverbial woods. No wonder, either, that the elements that comprise a tree's visual splendor—its dizzying proportions, its reassuring stasis, its branches expressing the duality of chaos and order—typically escape the attention of a race of puny, scurrying, way-too-civilized Homo sapiens.

Take the tallest and most massive trees of all—the redwood and sequoia. These paragons of natural selection embody a height and breadth and beauty so outsized as to defy mortal observation. Yes, we can fly over redwoods, which tower 360 feet or more, and regard their verdant canopies. Or we can stand at their Bunyanesque trunks and crane our necks, peering up to watch them foreshorten above us. But we can never quite encounter them as they are, all at one go. Enveloped by the branches of their neighbors, these green giants are visually inaccessible: human beings have never been able to step back, size them up, and admire their grandeur in full.

That is, until now.

As is evident in these pages, photographer James Balog has devised an inspired method for photographically isolating trees, top to bottom, limb to limb, and taking their complete graphic measure. The result is something of a milestone—aesthetically, conceptually, and environmentally. Balog has revealed to the human eye, for the first time, The Tree Itself, as creation made it. In so doing, he has forced us to stop, gape, and understand: in their looming majesty, trees not only command our attention, they demand it. Their monumental scale provokes awe. Their essential beauty beckons and inspires. Their sustaining life force shimmers in every leaf. If we ignore them, Balog's images tell us, we are ignoring the deeper beauties of life itself.

JAMES BALOG IS A NEO-NATURALIST. In the 1980s, he decided to create a new way of looking at and thinking about endangered animals and their vanishing habitats. He treated fauna in the same way that his peers treat fashion models. In a series of breakthrough photo sessions, he placed a mandrill, a jaguar, a Florida panther, and scores of other animals in front of a white backdrop, bathing them in artificial light. This unique styling, he now says, "gave me a doorway into their individual personalities, as living beings, that, for me, was new and intense and unexpected." The results were stunning. His images were crisply composed yet haunting, imbued with a hyperrealistic patina, as if he were snatching icons from some dreamscape out of Carl Jung, Salvador Dali, or Helmut Newton. The pictures were printed around the world—even on U.S. postage stamps—helping to raise funds for the endangered species movement.

Balog is also something of a visionary. In 1998, he realized that by summoning the same passion he had tapped into while depicting wildlife, he might be able to capture a corresponding dignity in America's largest trees. First he would need a method for rendering that grandeur on film. To that end, he tracked down his models: particular trees designated as the tallest of their kind. Next, he gathered huge white backdrops and flowing sheets of fabric the size of theater curtains, loaded them into an off-road vehicle, and set out for the forest. Once on site, he rented cranes and hoisting equipment and set out to photograph each tree as if in a studio, converting the great outdoors into his own diorama.

Yet while many of the pictures he produced were breathtaking, he wasn't satisfied artistically. The single frame didn't properly impart, in his words, "the expansiveness that the trees demanded." To meet his requirements for a more sweeping treatment, he next tried photo pairings, then triptychs and multipanel sequences. In one instance, he stood at the base of the country's most massive American elm, in Buckley, Michigan, and shot *outward*, toward the surrounding farmland, creating a 270-degree panorama, eight frames across, recording, in essence, what the *tree* sees every day. Still, these mini-frescoes didn't quite approximate the awe invoked by the spectacular dimensions of certain trees. What he needed, he realized, was a canvas vast enough to suit his subject and equipment dynamic enough to adapt to the ever-changing conditions in which he was photographing.

Finally, Balog settled on a truly innovative format: digital multiple-exposure, borrowing from the cubist sensibility of Picasso and Braque and building upon the mosaic-assemblage style pioneered by various photographic forebears, from imaging specialists at NASA to artist David Hockney. Balog—renowned as an adventure photographer and accomplished as a mountaineer who had climbed in the Himalayas, Alaska, and the Alps—understood that the best way to convey a tree's scale was to actually *scale* it. Soon, he was perched beside his subjects, within their branches' grasp, as a sure-eyed owl might view them.

Balog concocted a system involving a crossbow, a series of ropes, and a harness. He would first shoot an arrow at the tip of the tree he hoped to photograph, then secure a sturdy line across to a second tree opposite his target. Dangling several hundred feet in the air and sitting in a specially designed cradle, he would make his way down the length of the tree, using equipment developed by rock climbers and cavers. As he descended, he photographed at successively lower elevations, exposing a dozen or so images across each limb span, making scores of lateral passes as he dropped vertically, branch by painstaking branch. Certain trees would take him several hours to shoot. Winds would whip him to and fro in the treetops. He would ultimately land earthward, where his legs, like those of a collapsed marionette, would barely support him.

Once back home, in Boulder, Colorado, Balog would bivouac at his computer and spend as long as a fortnight on each mosaic, combining his airborne puzzle pieces into a single, fully realized image of the tree in toto—a thing of beauty on a titanic scale. "In the tangle of chaos," Balog says of his time spent hanging in the emerald lurch, "there is an overarching beauty that the mind doesn't immediately see in the wilderness of branches." But beauty he did forge, ingeniously, methodically, cutting a swath through the thicket that shrouds our everyday experience and creating a thoroughly fresh vision. His pointillist, serrated creations would mimic the organic, step-by-step growth process. The individual photo shards—as many as eight hundred per assemblage—would double as leaves on his digitally reconstructed tree. His final images, sometimes printed mural-size, would command the walls they graced. "Many of the tree portraits," he says, "had an uncanny way of spontaneously bursting into existence, as if they already existed somewhere and were just waiting for the chance to become manifest. More than once the trees seemed to be the creators, while my cameras were the vehicles bringing their intentions into being."

Balog's studies of trees are wondrous and curious artifacts: the first glimpse of creation's grande dames, viewed as a simulated whole. The photographs, says nature writer Gary Ferguson, author of *The Sylvan Path: A Journey Through America's Forests*, succeed in "lifting these magnificent elders from their surroundings and capturing their singular beauty, tree by tree. They remind us what an astonishing amount of life—in some cases entire civilizations—has come and gone while these trees alone endured."

James Balog has brought us up there, into the boughs, to dangle with him out on those limbs. And with him we soar to the limits of perception.

INTRODUCTION

GREAT TREES ARE SCULPTURALLY ELEGANT. They transcend time. They are humbling. They are authentic. They are nature's ultimate survivors, having escaped the ravages of weather, fire, disease, insects, and humans. They are even an antidote to the amnesia sweeping from one human generation to the next as we forget what makes nature natural. They remind us how the landscape once looked—and how it might look again if only humans had the desire.

From California to Ireland, from Iraq to China, the earth's original tree cover has been significantly altered or completely annihilated. Areas we now think of as naturally treeless—the upper Midwest of America and large parts of the Mediterranean basin, for example—were once blanketed by thick foliage. If photographing trees has taught me anything, it is this: age matters. Ancient forest is to regrowth as the Grand Canyon is to a ditch. Old forests have character, the imprint of time, biologic complexity, and architectural eloquence. Regrowth doesn't. On the ground, the difference between old forest and new is obvious to the eye. Beauty is something you can see and smell and touch and hear. It is precise, immediate, concrete. A response to aesthetics should guide a sane society and not be an afterthought; a society that forgets to see and love beauty is a hardscrabble place, parched to the bone. The United States has the wherewithal, and could reasonably be expected to have the intelligence and sensitivity, to do better.

Between 1998 and 2004, I went on a quest to photograph the largest, oldest, strongest trees in America. I was looking for stellar individuals, often referred to as "champions," recommended by arboreal aficionados or listed on the National Register of Big Trees. From Key West to the Pacific Northwest, from Maui to New England, the miles traveled became uncountable. My approach to making the images was unorthodox. Nature photography is fenced in by a century's worth of tradition and formula. Customarily, when a photographer makes an image of, say, a redwood, we see either the base of the trunk merging with lush groundcover or a column of wood tapering upward into the mist. With good light, proper technique, and some luck, the photograph will be an easily understood celebration of the tree. It will also be fundamentally identical to many of the redwood photos that came before. No matter how attractive such pictures might be or how much they might be a persuasive testament on behalf of nature, I was interested in finding something new.

A lifetime of visual influences helped shape my approach: religious iconography from the Middle Ages; painters like Picasso, Georges Braque, and Jackson Pollock; photographers as disparate as Vittorio Sella, Richard Avedon, Irving Penn, and David Hockney; and the photographic composites NASA produced when space probes and astronauts landed on the moon. My previous work with wildlife, where I was astounded to discover how much individuality each animal had, was even more vital in shaping the portfolio, though in the end, experience with the trees molded the photographs more than any other single factor.

Initially, I built enormous portrait studios in the forest, hanging artificial backgrounds behind the trees and lighting them with strobe lights. Supporting the backgrounds soon became impossible, at least given the limits of available funding. In addition, I noticed that light, subject, moment, weather, scale, mood, and a hundred other variables always seemed in flux. An eclectic range of visual treatments, it seemed, would better reflect chameleon reality. So I photographed in color one day, black-and-white the next, in formats ranging from 35mm to 4-by-5. Some images were complicated productions recorded on the highest-tech digital equipment, while others were snapshots on ultra low-tech plastic cameras. It took hundreds of days at the computer or in the darkroom to bring the images to fruition. Eventually, ninety-two individual trees, of forty-seven different species, made their way into this book.

America's master narrative claims that the continent was pristine wilderness before European colonists came. In truth, something like 10 to 20 million Native Americans were already here. They cleared trees to plant crops. They burned enormous amounts of firewood for heating and cooking. They torched forests to maintain grassy openings in woodlands and foster populations of deer, elk, and bison (incredible as it now seems, bison once roamed as far east as Pittsburgh). Burning shifted the ecology toward nut-bearing species like oak, chestnut, and hickory, providing food for animals and people alike. The pre-Columbian wilderness was thus not immutable and pristine at all. Still, with relatively modest populations and limited technology, Native Americans hit the forests with far less force than Europeans, with their more invasive farming techniques and industrial-grade agriculture, soon did.

The first colonists built their cabins at Jamestown, Virginia, in 1607 and Plymouth, Massachusetts, in 1620. These settlers, and those who followed, mowed down trees as energetically as any farmer in Amazonia does today. The eastern hardwood forest, which once stretched from the Atlantic Ocean to Iowa and Oklahoma, metamorphosed into farmland; the trees became building materials, firewood, and charcoal.

By 1850, 100 million acres of primordial hardwood had vanished. Between 1850 and 1860, another 10 million disappeared. The great pine woods of Michigan, Wisconsin, and Minnesota were leveled in the latter decades of the nineteenth century. The vaulted conifers of California and the Pacific Northwest hit the ground starting around 1850; the advent of chainsaws and logging trucks in the early twentieth century, and then the post–World War II housing boom, greatly sped up the process.

Today, in the deciduous forest of central and eastern North America, primeval, virgin woodland occupies half of 1 percent of the 350 million or so acres that are presently forested. In the lush lowlands of the Pacific Northwest, where the biggest conifers in the world grow, the story is the same: 99.5 percent of the ancient forests are gone. In the lower forty-eight states, only 5 or 6 percent of forest is virgin today; most of that is in the high montane areas of the Rockies, Sierra Nevada, and Cascades. America's last extensive stands of primeval lowland conifers are in the Alaska panhandle, and most of those will be finished off in another decade or two.

With or without human intervention, the natural world is always in flux, of course. Since the first, large-format edition of this book was published one short year ago, much has happened. Nameless seedlings, some perhaps destined to be future champions, sprout from one coast to another. The tallest tree in the world, a redwood named Stratosphere Giant, grew from 369 feet $10^{1}/2$ inches to 370 feet 2 inches. The General Washington giant sequoia, once the second largest tree (by wood volume) in the world, became a shadow of its former self: a lightning-caused fire in 2003 had burned it from a height of 254 feet down to 229 feet, but winter storms in January 2005 snapped the trunk off at just 115 feet. America's champion elm is essentially dead. The inorganic world is also in flux, creating its own impact on plant life. Mount Saint Helens erupted again. The subducting tectonic plates of the Sunda Trench produced an epic earthquake, spawning the great Indian Ocean tsunami of December 26, 2004. And this, in turn, leads to a story about a certain special tree.

MY ENTIRE ADULT LIFE HAS, IN ONE WAY OR another, been driven by a desire to bear witness to the forces of nature. After the tsunami, I was obsessed with going to Banda Aceh, on the west coast of the Indonesian island of Sumatra, where the tsunami had hit hardest; after all, this was the most powerful natural event that would likely happen in my lifetime, and I needed to see it. I had already experienced firsthand what huge hurricanes, volcanoes, avalanches, and floods can do, and had seen the barrage of images and stories about the tsunami in the news. I was hardly naïve about what Banda Aceh had in store. But photographing human disaster was not my core objective. For one thing, the news photographers preceding me had already done excellent work covering the human toll. More importantly, the philosophy so many of us live by maintains that nature is inherently good. Though from a human perspective the tsunami was a catastrophe, it was more: a manifestation of the essential churning of immortal geologic time, without malice toward humanity at all. My pictures would somehow, I hoped, emphasize not disaster but power, not destruction but life.

The real world is never as simple as the abstractions of philosophy. The scope and severity of the tsunami's destruction were far worse than

anything I had ever seen or imagined. Contrary to assumptions made by many Americans—and I sheepishly include myself in that number—Banda Aceh had not been a destitute city of ramshackle fishing shacks. Rather, its structures were built so well of masonry, concrete, and iron rebar that the second largest earthquake in recorded history, centered just a few dozen miles away, had barely affected it. But the tsunami was far more powerful than the quake. Nothing could resist its mountains of water until they had avalanched so far inland their own internal physics collapsed. As I wandered through the impact zone, a fetid odor—the combined stench of ocean-soaked mud, furniture, lumber, plants, food, clothing, books, cars, computers, and every other object known to civilization—hung over the flattened city. I quickly learned to distinguish between that generalized smell and the specific one of decaying humans. Sometimes I abandoned photography to participate in retrieving and burying corpses. Repeatedly, the thought flashed through my mind: so this, *this* is what it will look like when the world as we know it comes to an end. If my years of wildlife photography had made me patient and tree photography had made me humble, then tsunami photography made me feel as ephemeral and fragile as a snowflake.

The incoming waves hit the northwestern fringe of Banda Aceh first, along the shoreline of a neighborhood called Ulee Lheu (pronounced "oo-lay'-lee"). The shocking satellite photos showing Banda Aceh before and after the tsunami, widely disseminated on television and in print, were taken looking straight down on the Ulee Lheu beachfront. Before the waves hit, the community had large, elegant stucco houses with red tile roofs, homes that would fit in among the opulent residences of Santa Barbara or Palm Beach. The ocean reduced every single one of them to rubble, much of it small enough to fit in an ordinary trash can. Only a few shattered walls still stood knee- or chest-high. A large slice of land had washed out to sea. Yet the place held a magnetic allure. The ruins were the only place in the city where large numbers of Acehnese would lounge around outdoors. At sunset, dozens of people wandered around, chatting and throwing stones into the sea. Mostly they just gazed contemplatively across the silver abyss.

Nothing along the beachfront survived the wave, with one mind-boggling exception: a tree. I wish I knew what kind of tree it was, but no one could tell me. Its feathery foliage resembled a cypress. Its trunk was five or six feet in diameter. With nothing standing around the tree, its height was hard to estimate, but it was at least 100 feet tall, and possibly 125 feet. Its smooth bark, much like that of a beech, had been scrubbed to a pale gray up to a point roughly 45 feet above sea level. This was the high water mark of the aqueous monster that had surged past it (the foliage showed signs of having been splashed by salt water nearly 20 feet higher). The tree's endurance in the devastated landscape seemed to have a lot to do with why people congregated at this spot. I photographed this arboreal marvel many times.

When I returned home, my heart was bled white by exhaustion and horror. Treading such hallowed ground and getting a glimpse into the immensity of geologic time left a feeling of having had a religious experience. Yet there was an undertone of frustration from never having found a picture to symbolize the pure power of nature. How could I, when the roaring waves had come and gone before I ever arrived on Sumatra, and only secondary effects like destroyed houses remained to be seen? It took some months to realize the image I wanted was there after all. The tree towering amid the wreckage of Ulee Lheu was the power itself: life force had outlasted geologic force. I know now that at the end of the world a tree will still be standing.

COAST
REDWOOD
"El Viejo del Norte"
Jedediah Smith Redwoods
State Park, California

TEXAS LIVE OAK

Quercus virginiana var. *fusiformis*

Rio Frio, Texas

This national champion lives in a tranquil valley deep in the Texas Hill Country—a down-home kind of place where women are regularly addressed as "ma'am," where bass glide through emerald pools, and where, during hunting season, you can see men skinning whitetail deer along the main street of a small town.

The oak—trunk circumference 295 inches and crown spread 95 feet—is thought to be more than nine hundred years old. Surveyors used it as a bearing point when the hamlet of Rio Frio was platted late in the 1800s. Today, the oak grows in the front yard of a charming bed-and-breakfast.

The fabric was originally meant as a background for isolating the tree's form. At one point, a breeze swirled through. When I darted behind the scrim to secure some ropes, I noticed an evocative shadow projected onto it. We quickly rearranged lights and camera. As the exposures clicked through the camera, a curious feline made a sudden appearance, then was gone.

Above and opposite:

TEXAS LIVE OAK

Rio Frio, Texas

FREMONT COTTONWOOD

Populus fremontii var. *fremontii*

Patagonia, Arizona

West of the 100th meridian, the longitude of western Kansas and the center of the Dakotas, aridity makes life hard on trees. Mountains may catch snow and nurture conifers. Artificial conditions might nurture exotic species in cities, towns, and backyards. But under natural conditions, only cottonwood survive in the desiccated plains. Under the cerulean sky in places like the Nevada desert or Wyoming prairie, where nothing else grows taller than waist level, fringes of cottonwood can be visible from 10 miles away, marking seams of water stitched through the earth.

It still came as a considerable surprise to find this national champion growing an hour-and-a-half's drive south of Tucson, a few miles away from the Mexican border. I had fully expected to see the landscape becoming ever drier and more filled with cactus. Instead, a lush upland appeared, complete with a charming little hamlet named after southernmost Argentina. Nearby, along Sonoita Creek, grew this enormous vegetable. The cottonwood's trunk, at 42 feet in circumference, is as big around as a respectable redwood. It stands nine stories tall and its crown is more than 100 feet wide.

Above and opposite:

FREMONT COTTONWOOD

Patagonia, Arizona

FREMONT COTTONWOOD

Patagonia, Arizona

By the time the Pilgrims landed in the *Mayflower*, the Wye Oak was already as large as most of the oaks we see in America today. Eventually, it became the centerpiece of a 5-acre state park, reigning as the national champion since the big-tree register was established in 1940. It grew to be 384 inches in trunk circumference and 96 feet tall, with a crown spread of 119 feet.

Major limbs fell off the tree in 1953, 1956, and 1984. To prevent further destruction, arborists surrounded it with a fence, packed plastic and concrete fillings in trunk cavities, installed three lightning rods, and wove a support armature of $1^{1}/2$ miles of steel cable through its elephantine branches. I visited the Wye Oak, hoping to photograph that armature. But masses of leathery, autumn-browned leaves still hung on the oak, making the cable hard to see. So I made an image that placed the tree in the context of the modern landscape: grain storage silos on one side, manicured exotic plants on the other.

I intended to return someday when the leaves had fully fallen, but circumstance intervened. On June 6, 2002, the Wye Oak blew down in a violent thunderstorm. It was more than 460 years old. Some leaves were cast in copper and turned into artworks, and its obdurate wood was milled into a new desk for the governor of Maryland.

34

SASSAFRAS

Sassafras albidum

Maggie Valley,
North Carolina

During the soporific droning of elementary-school science classes, I learned that photosynthesis and chlorophyll were topics best avoided. Happily, travels with trees reeducated me.

Chlorophyll and blood are nearly identical substances. They differ by only a single atom out of the 137 atoms in each of their molecular structures: chlorophyll contains an atom of magnesium while blood has one of iron (the rest are carbon, hydrogen, nitrogen, and oxygen). The entire terrestrial food chain, from redwood and prairie grass to elephants and hummingbirds, depends on the mineral distinction of chlorophyll. It captures the red and orange rays of sunlight raining upon the earth, mixes them with water and air, and creates the pure energy of glucose. The process is, of course, photosynthesis.

The chlorophyll-rich solar collectors we call leaves unfold remarkably quickly in springtime if one watches carefully. Up to 100,000 leaves might grow on an apple tree, 200,000 on an average birch, and 700,000 on a large oak. On North Carolina's largest sassafras, no leaves were visible one day, but a verdant halo burst out the next.

YELLOW-POPLAR
(a.k.a. **TULIPTREE**)
Liriodendron tulipifera
Great Smoky Mountains,
North Carolina

One Easter Sunday, I rappelled through a grove of ancient trees in the southern Smokies. Yellow-poplar trunks 9 feet in diameter towered in the cathedral-like forest. Their bark looked as durable as marble, yet to the touch, it was as airy as balsa wood. Chlorophyll flooded vibrant green into billions of rambunctious young leaves. Ivory-colored dogwood and silverbell blossoms laced the understory. Delicate little plants that would be right at home in a fable about leprechauns carpeted the ground. Three-inch-thick moss coated the rocks. The calls of a dozen different bird species echoed; one sounded like a manic gibbon.

Two-thirds of the way through my rappel, a thunderstorm hurtled over the mountains. The lightning came much too close, and juicy raindrops spattered the camera. Haste became the new order of the day. I gave up photographing from sequential positions and covered the remaining woodland from a single point on the rope. On the final frame, I reached out and held a leaf so fresh that it seemed to be unfurling before my eyes.

WHITE OAK

Quercus alba

Granby, Connecticut

Lightning strikes oak more often than it does other species. To ancient cultures, this made oak appear to be conduits for bringing the power of the sky gods down to earth. Accordingly, the trees were held in great reverence. In our scientific age, we have other theories: lightning is thought to be magnetically attracted by the damp wood in the core of ancient, decaying oak. Or, possibly, during rainstorms the rough, absorbent bark allows oak to become more saturated than other trees, drawing lightning the way open water does.

When Connecticut was first colonized, a magnificent oak stood where the city of Hartford would soon rise. In 1637, a Native American chief named Sequassen begged the settlers not to cut it because, according to one account, it had "guided our ancestors for hundreds of years, telling us where to plant the corn. When the leaves are the size of a mouse's ear, then it is time to put the seed in the ground." During a 1687 legal spat with the king of England, the colony's charter was hidden in the oak's hollow trunk. The tree became known as the Charter Oak. It was blown down in 1856. The closest approximation of the Charter Oak today is this exquisite tree in Granby, which sprouted around the time of English colonization.

WHITE OAK
Granby, Connecticut

EASTERN WHITE PINE

Pinus strobus

Lenox, Massachusetts

Tree devotee Bob Leverett generously tipped me off to this pine's existence. The Eastern Native Tree Society, the premier association of those enraptured with ancient deciduous trees, was founded in Bob's kitchen. In the past two decades, he has delineated fifty-five previously unknown patches of old-growth and measured some six thousand individual trees.

Big trees give a fine excuse to go wandering. This conifer is deep in an ordinary-looking forest called Bullard Woods, $1/2$ mile away from the grounds of the famous Tanglewood Music Festival. Leverett believes that the tree germinated around 1820, just before Massachusetts had reached its point of maximum deforestation, when something like 50 to 60 percent of the land was treeless. At that time, the pine was surrounded by open pasture. The birch and red maple seen around it are still in the process of filling in the old pasture. Look carefully and you can see three people at ground level.

Early on a rainy winter morning, it was still dark as night. My assistant, John Wiltse, and I left Olympic National Park and splashed up U.S. 101 along the seaward edge of the Olympic Peninsula. Our objective was a gigantic cedar in a different section of the park. But before we could get to it, we had to cross a profound perceptual divide: we exited preserved parkland and entered logging country.

At 5:45 A.M., we pulled off the highway for breakfast at a little café named JJ's. A blue fog of cigarette smoke floated above a Formica countertop so old and so scoured that an underlayer of blackened wood showed through. A picture of a spotted owl graced the wall. Below it a bumper sticker read: "Are you an environmentalist? Or do you work for a living?" At the counter sat two loggers in flannel shirts and mud-caked canvas work pants. Neither had come within arm's reach of a razor in days. Their creased faces had the greenish white pallor acquired from decades of inhaling diesel and hydraulic fumes. They smoked Camel cigarettes and sipped tarry coffee as they recited a litany of complaints the tired waitress must have heard countless times before. A compressor had broken down. Replacement parts were impossible to find. Repairs were improvised, probably inadequate. John and I quickly mopped up our maple syrup, pancakes, and eggs without joining the discussion. As we left, I noticed another bumper sticker on the wall: "Earth first. We'll log the other planets later."

We kept heading north. Empty log trucks roared past us every few minutes as they headed to the clear-cuts for their first loads. They would soon return stacked with relatively spindly logs, since the big trees had been cut by the fathers and grandfathers of the men working the woods today and processed into lumber for American homes. (Northwest Coast Indians once made houses, totem poles, and canoes from cedar, and baskets and clothing from the bark.) As day broke, razored hills appeared. Eventually, we reentered the park and found the Kalaloch Cedar.

The tree had been the national champion from 1955 until 1977, when the Nolan Creek Cedar was discovered. Given the density of the hobbitlike rain forest, a clear view of this fantastically gnarled organism was impossible. I slowly orbited the 19-foot-diameter trunk and photographed sections as I went. At one point I tried to get a location fix with my GPS, but in the profusion of the forest, I couldn't acquire a satellite signal. The display screen asked me to identify whether I was in Afghanistan or Malaysia or Zimbabwe. I was reassured to know that at least some places on this earth are invisible to the roving digital eyes overhead.

Above and opposite:
WESTERN REDCEDAR
"Kalaloch Cedar"
Olympic National Park, Washington

Above and opposite:
WESTERN REDCEDAR
"Kalaloch Cedar"
Olympic National Park, Washington

The General Sherman is the world's largest single living organism. Though its 274-foot height makes it nearly 100 feet shorter than the tallest redwood, the tremendous bulk of its trunk—27.1 feet in diameter, 102 feet in circumference at ground level—puts it in the record books with 55,040 cubic feet of living wood. The Sherm, as one tree-loving friend of mine calls it, adds about $1/32$ inch of diameter every year. Its precise age is unknown, but 2,100 to 2,200 years is the current best guess. The species lives at altitudes between 5,000 and 8,000 feet on the western flank of the Sierra Nevada. I focused my photography on the stupendous junction between the tree's life and the earth that nourishes it.

The existence of this great species first became widely known during the mid-nineteenth century. In 1852, a man named Augustus Dowd, who was hunting grizzly bear, stumbled onto a giant tree he believed was the biggest in the world. In celebration of its splendor, the tree was cut down and a portion of it shipped to the East Coast for display.

By 1879, the Sherm had been named after Union general William Tecumseh Sherman, an act of considerable irony: the stern-fisted general incinerated more lumber than anyone in the history of North America during his 1864 March to the Sea, and he went on to command the war that exterminated the free-roaming Plains Indians. Late in the 1800s, socialists living in a mountain commune called the Kaweah Colony renamed the tree after their own hero, Karl Marx. The general's name was subsequently reinstated. Recently, a group of horticultural entrepreneurs has been attempting to clone the Sherm and, without any particular legal, financial, or moral authority to do so, sell naming rights to the offspring.

Debate long raged over which California county could claim the world's largest woody object. During the 1920s, a team of surveyors made careful measurements of trees known today as Boole, Hart, General Grant, and General Sherman. Sherman won easily. For the past several decades, a small cadre of devotees, led originally by the late Wendell Flint, hiked every nook and cranny of the Sierra looking for rivals. None has topped the Sherm, nor has any of them surpassed the other trees long accepted to be in the top six. But to this day, the identity and relative ranking of the rest of the top twenty is in flux; it is as if the Himalayas were still being surveyed and the identity of the 8,000-meter peaks remained uncertain. In mid-September 2003, a forest fire burned the crown off General Washington, long considered the world's second largest tree. This, of course, changed the relative ranking of every tree except the Sherm.

VALLEY OAK

Quercus lobata

Covelo, California

If, as Wallace Stegner said, "Space is a place with no memory," then a great tree infuses empty space with memory and turns it into a place, creating a bridge between civilization and wildest wilderness.

This valley oak germinated long before Anglos started farming a tranquil isolated basin in the northern California hills. Many no doubt could have found reason enough to cut down the tree, but didn't. Its current owners, Bobby and Sheila Fetzer, members of a noted viniculture family, are enormously proud of their national champion.

As I made the first exposures, I was dangling only an arm's length from the branch visible on the top left side of the oak. The circular shape on the limb marks a hollow where a colony of bees lived. They buzzed in and out of their hive just a few feet from my face. I had two choices: go down and try a different image—or take a deep breath, think friendly thoughts, and hope the bees' intentions were amicable. They were.

VALLEY OAK
Covelo, California

VALLEY OAK
Covelo, California

SITKA SPRUCE
"Cape Meares Giant"
Cape Meares State Park, Oregon

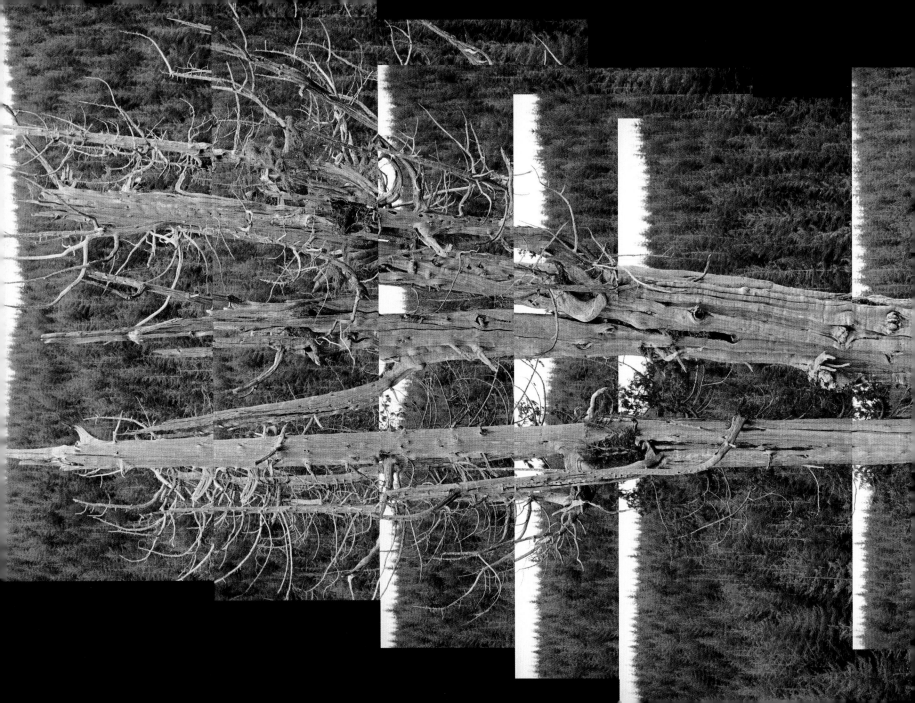

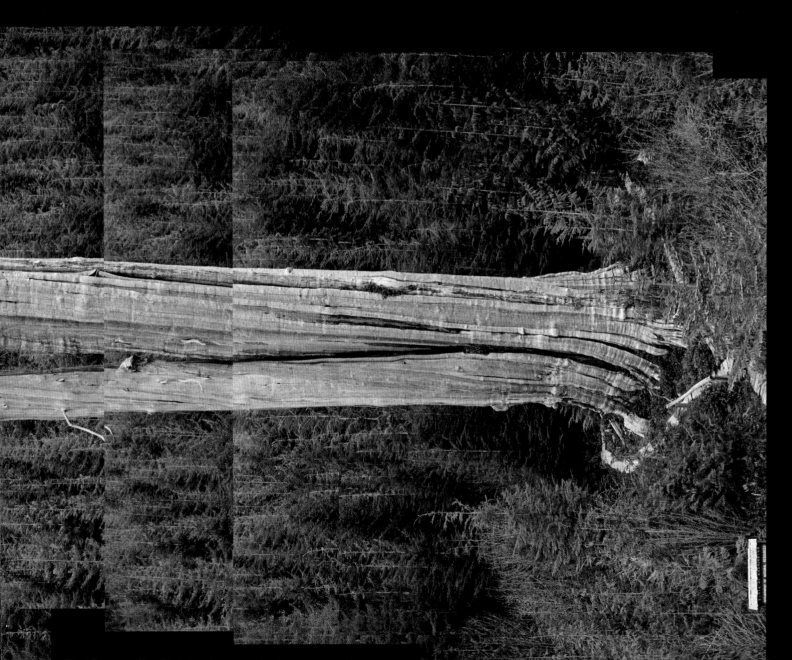

WESTERN REDCEDAR
"Nolan Creek Cedar"
Olympic Peninsula, Washington

WESTERN REDCEDAR

Thuja plicata

"Nolan Creek Cedar"

Olympic Peninsula, Washington

The lushest forests in the lower forty-eight states are found on the western side of Washington's Olympic Peninsula, nourished by as much as 10 to 20 feet of annual rainfall. In fact, the tallest tree ever measured in North America, a 385-foot-tall Douglas fir, grew on the peninsula (it was cut in the nineteenth century). Here, redcedar, Douglas fir, and Sitka spruce reach truly epic proportions. Olympic National Park still preserves pockets of virgin forest, but most of the primeval trees have long since been turned into lumber. In the late 1970s, logging crews were clear-cutting a district called Nolan Creek, owned by the state's Department of Natural Resources. This cedar was spared when it turned out to be a national champion. Today 30-foot-tall regrowth surrounds it.

The Nolan Creek Cedar is 18.8 feet in diameter, 171 feet tall, and roughly two thousand years old. Living bark covers just a small section of trunk, yet continues to nourish living branches high in the crown. The rest of the trunk has been weather-beaten to a ghostly gray. I wanted to put the cedar in the context of the clear-cut around it but could find no satisfactory earthbound vantage point. Phone calls revealed that a helicopter was working with a logging crew a few miles away. As soon as the pilot finished his work there, he zoomed over and picked me up on a dirt road. We rotored up and down repeatedly, working out different visual treatments. The emptiness of the sky evokes the vanished arboreal lives.

WESTERN REDCEDAR
"Nolan Creek Cedar"
Olympic Peninsula, Washington

WESTERN REDCEDAR
"Nolan Creek Cedar"
Olympic Peninsula, Washington

69

AMERICAN ELM
Buckley, Michigan

Rural Michigan is charming in a classically agrarian way. Barns and silos, snug farmhouses, and neat rows of corn all speak of an earnest and prosperous attempt to wrest a living from native soil. The area's ecological history, however, is for the most part invisible.

Until relatively late in the settlement of the Midwest, this land was covered with a thick forest of white pine. The currents of America's Manifest Destiny flowed mainly westward, with few settlers making the turn north into the Michigan woods. But after the Civil War came an explosive demand for lumber to build cities; by the 1890s, logging had largely stripped the land of pines.

For many years, an elm named the Louis Vieux Tree was the national champion. It grew along the Vermillion River in northeastern Kansas. In 1997, a young man firebombed the tree. When the police arrived, flames were shooting 25 feet into the air. He was caught, fined, and sentenced to probation—which included planting an elm sapling.

The tree pictured here became the new champion. More than 1,800 visitors entered their names in a log book near the elm during the four years preceding my arrival in 2000. The tree unfortunately has Dutch elm disease, the beetle-borne blight first imported on timber from Europe in 1909, and many of its roots were inadvertently cut by an overzealous farmer plowing his fields. As of this writing, the tree is nearly dead.

Photographing such a graceful tree in a single frame seemed to confine it. Instead, I let the branches extending into the sky suggest the composition of the triptych on the previous pages. The seven-panel image above shows the tree's view, looking north, east, and south across the farmland. A raw November breeze whistled through the branches as I photographed. Simultaneously, votes in the Bush-Gore race for president were being cast. Today's politics were shaping tomorrow's ecological history.

Left and opposite:
AMERICAN ELM
Buckley, Michigan

74

AMERICAN BEECH

Fagus grandifolia

Waynesville, North Carolina

This beech was surrounded by some of the most abject squalor I have seen in an economically developed country. The rotting carcasses of long-dead cars, rusted appliances, and scrap metal were thrown in chaotic heaps. The inhabitant of a collapsing single-wide house trailer muttered incoherently. A dozen bloated hogs lived in a miserable pen. A vile lagoon held their effluent, and the odor was unspeakable.

Paradoxically, an unpleasant situation led to a satisfying picture. The hogs, it seems, had been rooting around the beech, eroding the soil and exposing the tree's subterranean plumbing. Seeing the hidden underworld of a living tree is extremely rare, so I decided to make the picture on the opposite page even though the beech was not a champion.

A common misconception is that the mass of a tree's branches mirrors the mass of its roots. Rarely is this true, but the aboveground and belowground worlds obviously have similar morphology. By leaving out certain frames from the finished work, I took the idea of reflection a step further: the shape of the image evoked the shape of the subject. The beech on the following pages is the national champion.

Above and opposite:
AMERICAN BEECH
Lothian, Maryland

Photography is a ritual of engaging with the gift of sight. One learns to see, not just look. The generalized becomes particular and intimate, alive with meaning and memory. For me, few trees evoke such experience as well as this spicebush.

Though cochampion of its species, the delicate little tree is only 23 feet tall and 14 inches in circumference (a plant qualifies as a tree if its trunk is at least 3 inches in diameter and 4 1/2 feet above the ground, with a crown of foliage 13 feet tall or more). At first light one autumn morning, tree aficionado Mike Davie led my assistant and me on a mile-long hike down the Swannanoa River floodplain to see it. Temperatures the night before had been below freezing: thick frost crackled on the leaves underfoot. As soon as we strung the background fabric into the surrounding trees, the form of the spicebush was liberated from the thicket behind it. The air was utterly immobile, the forest still and silent.

At the end of every leaf stem, or petiole, is a section of cells called the abscission zone. When water in the abscission zone freezes, as it had the previous night, it severs leaf from twig. For a leaf to drop requires only that the ice in the abscission zone thaw or that a breeze break it off.

Around 7:30, the sun started inching up in the east. Sunbeams slanted through the thick river-bottom vegetation. The air warmed and petioles thawed. At 8:10, a leaf suddenly fell off the spicebush. Another dropped, then another. Within minutes, leaves were fluttering around us as if the forest had been invaded by a flock of butterflies. By 8:50, the branches were all but naked. Never before had I understood so clearly why autumn is also called the fall.

The frame shown here was exposed around 8:25. In hindsight, I regret not having a sequence to show the transition. But I couldn't have guessed how radically the tree would transform itself in such a short time. The memory of the fall makes the image a rich one for me still. In addition, this photograph marked the beginning of learning to see rhythm and order within apparently chaotic scenes.

REDBUD

Cercis canadensis var. *canadensis*

Maggie Valley, North Carolina

The grace of blossoming trees has been widely celebrated in both Eastern and Western art. No matter one's age or cultural background, the elegance of a blossoming springtime forest is impossible to resist. If redwood and sequoia are imbued with an inherently masculine quality, flowering trees possess an allure not unlike the curves of the female body or the elegant lines of haute couture.

One springtime found me in the Great Smoky Mountains, with flowering trees projecting their radiance everywhere. Growing outside the small cabin I called home for a couple weeks was this gorgeous redbud. I tried to defy its siren song—after all it was not a champion, and there were plenty of those to chase—but its charms grew too strong to resist. One day, I gave in and made this portrait.

Sycamore are the largest hardwood in America. Stupendous individuals, some five hundred years old and 40 feet in circumference, were once common. After sycamore reach their third century, the trunks become hollow, making them prime habitat for all sorts of fauna. Two young men reputedly made their home inside an old West Virginia sycamore for three years during the pioneer era.

I happened upon this North Carolina state champion one spring. The spheres resembling Christmas tree ornaments are seedpods waiting to drop. I was entranced by the way the pods and branches created a harmonious cadence even though the arrangement didn't fit conventional ideas about proper composition. By this time, I'd come to understand how sylvan life manifested fractal geometry and chaos theory: the weaving patterns of branches repeat themselves at different scales of observation, from the distant to the microscopic, all of it permeated with a mathematical rhythm that is at the heart of even the most disorderly-seeming nature. Though the principle was first quantified by mathematician Benoit Mandelbrot in 1980, artists like Leonardo da Vinci, eighteenth-century Japanese painter Katsushika Hokusai, and twentieth-century Abstract Impressionist Jackson Pollock understood it intuitively.

Above and opposite:

SYCAMORE

Ironduff, North Carolina

AMERICAN CHESTNUT

Castanea dentata

Cicero, Washington

Chestnut were once widespread throughout the eastern United States. But in 1904, a fungal bark disease carried by nursery trees imported from Asia began to spread. Within less than a human generation, 9 million acres of chestnut were dead. The species was all but exterminated.

Large chestnut trees are rare today and can be found only outside their original home range, where the blight couldn't reach them. Following directions from Bob Van Pelt, one of the leading American authorities on big trees, I found myself driving one day up the Stillaguamish River valley of Washington State. The largest chestnut in America, he had told me, could be found on the edge of a pasture at mile marker 30, with a black walnut tree on one side and a butternut on the other. But when I arrived, the object of my desire had been blown down by a storm. Most of the chestnut had been sawed up and carted away. A remaining piece showed that the tree had grown at the phenomenal rate of $1/2$ inch per year. I counted sixty-five rings and estimated that the hollow core once contained another fifty to seventy. The tree had thus been planted sometime between 1870 and 1890, probably just after clear-cutting of the primordial forest. Van Pelt calls trees planted by the early settlers "pioneer trees."

I eventually found my way to two other chestnuts that the Cicero tree's demise had promoted to cochampion status. One shades a cemetery in Tumwater; the other grows on the Washington side of the Columbia River, between a bank parking lot and small apartment buildings in a gritty mill town called Washougal. Both appear to be pioneer trees, too.

AMERICAN CHESTNUT

Tumwater, Washington

AMERICAN CHESTNUT

Washougal, Washington

The Jack-and-the-beanstalk moment was upon me, the instant when some elemental part of my wiring—the survival-obsessed brain stem, I suppose—decided to stop fretting about how high above the ground I was. Once cut loose from its anxious shackles, my mind was free to float up into the giant's kingdom of the high canopy.

I spun on a rope through empty air ten stories above the snow-covered forest floor. All around, as far as I could see, rolling waves of khaki treetops mingled with rust-orange sequoia trunks. The Alder Creek basin, on the western flank of the Sierra Nevada, drops down toward the San Joaquin Valley and eventually the ocean. A slab of pale gray clouds sat thickly overhead, promising rain.

Sequoiadendron giganteum genes have given birth to all the largest trees in the world, as measured by the volume of their wood. A few dozen yards away from me towered one of the biggest of the big. During the past two millennia or so, this giant vegetable, the fifth largest tree in the world, has managed to grow 242 feet tall and 25 1/2 feet in diameter. Forty-four thousand one hundred cubic feet of arboreal life pullulates into the Sierra sky. The tree is named after Amos Alonzo Stagg, a noted college football coach of the 1950s and '60s. Stagg had little admiration for trees as far as anyone knows, but a civil engineer who built a road near the sequoia admired the coach. The engineer suggested the name to the Rouch family, who has owned the square mile of land around the tree since the late 1800s, and it stuck.

Being 100 feet above the ground on a rock climb, when body and mind cling to obdurate stone, never seems terribly high. But my situation here was different. Three small sliding clamps, known as jumars, attached me to a black 11-millimeter rope. This rope went another 150 feet straight into the sky, where it was tied to a line strung laterally (known as a Tyrolean traverse) from the top of Stagg to one of the neighboring trees. The whole system was at the mercy of atmospheric swells rolling in with a winter storm, and I bobbed up and down as they blew through. So it was a relief when my mind gave up its love of terra firma, accepted its fate, and let me climb to the sky in peace.

I chugged methodically along, sliding the ascenders 16 inches or so up the rope on each step. Since my left arm was already in a cast because of torn wrist ligaments (the product of a slight carpentry accident followed by difficulties rigging a stubborn tree a few weeks previously), and since mistakes with the equipment can be so difficult to unravel when dangling on a rope, I took great pains to advance with precision and care. I finally reached the top of the magic beanstalk after forty-seven minutes. Twenty-five stories above the earth, where the climbing rope clipped into the traverse rope, I couldn't help marveling at how strange was the compulsion that brought me to this solitary point in three-dimensional space; my life depended on two $16 loops of aluminum no thicker than my pinkie.

"You look completely insane up there, man," laughed Jim Spickler through the walkie-talkie looped around my neck. Spickler was one of two masterminds responsible for getting my ropes into Stagg. Billy Ellyson, the other rigger, worked his way up a rope tied to Stagg's crown. Their work was a marvel of engineering, adrenaline, and muscle. (Exactly how they did it I am under oath not to tell, and can say only that placing the lines had something to do with crossbows.) I wouldn't be here without them.

On this December day the thermometer was barely above freezing. Rain began; it went on without letup until nightfall, when it changed to heavy snow. I carefully clipped myself into a rappelling device, then unclipped my foot loops and chest harness from the rope. This moment of switching into rappel mode is the evil twin of the Jack-and-the-beanstalk moment; a mistake with knots or carabiners would be fatal. Fortunately, the moment passed in good order. I untangled my camera from the thicket of nylon slings around my neck, rearranged a five-gigabyte portable hard drive and two flash cards in the little pouches sewn to my harness, and was ready to begin.

My digital Nikon essentially acted as a freeze-frame motion picture camera. I panned horizontally, shooting eight frames from the left of Stagg all the way to the right. Then I rappelled down 15 feet and repeated the process. Over and over I descended in measured increments so that no twig went unrecorded. At times, the rope spun me in circles; it became impossible to know if I missed parts of the tree, so sometimes I covered the same pan more than once. After each flash card in the camera filled up, the images had to be uploaded into the portable hard drive. Simultaneously, I shot on a second card. Once it too was full, I erased the first card, put it back in the camera, and kept going. The original exposures thus were wiped away as I descended. I could only pray the electronics were working properly.

The rain thickened. Electronics got wetter. Icy moisture soaked me to the skin through a full-body Gore-Tex suit and two layers of fleece. When my boot soles touched ground, after nearly four hours in the air, my legs were numb, and for a few minutes I couldn't stand properly. Four hundred fifty-one frames had apparently been transferred to the hard drive. But not until I reached our cabin, long after we'd pulled the ropes from the sequoia and my rigging team had dispersed, did I know if the fickle electronic images had survived intact.

Happily, they did. Back in the studio a week later, my keyboard-tapping fingers digitally rebuilt Stagg. Stitching a reasonable semblance of the arboreal tapestry together required two weeks; perfecting it took four more. Watching the sequoia emerge was a revelation, since I had never actually seen the tree whole, but only in fragments as I climbed or rappelled through the forest. Ansel Adams is well known for having called his camera negatives the musical score and his prints the performance. I'll expand the metaphor further: if the individual exposures are isolated melodies, the intact composite is the complete symphony.

COAST REDWOOD

Sequoia sempervirens

"Iluvatar"

Prairie Creek
Redwoods State Park,
California

Iluvatar (pronounced "ill-oo´-va-tar") is the most distinguished individual in the coastal rain forest of Prairie Creek Redwoods State Park. It was named after the creator of the universe in J.R.R. Tolkien's *The Silmarillion*. Depending on which set of measurements and definitions one adheres to, Iluvatar may be either the largest or the second- or third-largest redwood by wood volume in the world, with roughly 36,470 cubic feet of wood perched on top of a base more than 20 feet in diameter. Iluvatar has an unquenchable lust for life. Bob Van Pelt believes it is the most architecturally complex tree in the world, with an upper trunk that breaks out into a candelabra of 134 lesser trunks called iterations.

If they were freestanding, many of these iterations would be considered enormous trees in their own right. The largest is 8.5 feet in diameter, and has a complete colony of ferns, orchids, salamanders, and lichens growing on top of it. Some iterations are so intertwined with others that they have fused together into an Eiffel Tower–like superstructure of reinforced branches.

Redwood grow to the sizes they do because of their physiology. Like cedar and juniper, they are rich in polyphenolic compounds, which color the wood red and repel insects and disease. Bark up to a foot thick insulates the trees from cyclical forest fires as effectively as luxuriant fur protects a polar bear from winter. Redwood have the largest wood cells, or tracheids, of any conifer. Their plumbing is uniquely effective at slurping water out of the ground and pumping it up to the highest branches: the process is known as transpiration, and it takes as much as twenty-four days for water to make its way from roots to the highest branches.

Caution to all: redwoods can be intoxicating, if not addicting. Scientist Steve Sillett had been an undergraduate biology major when, unroped and alone, he climbed up the furrows in the bark of a large redwood. When he discovered that the top of the tree had broken off and he couldn't get the expansive view he hoped for, he jumped, squirrel-style, across to a thin branch on an adjacent tree. "I would never do that again," Sillett says. "But I made it, and when I got to the top and saw that amazing view, I was hooked on big trees for good." At times, he can be a sober scientist, rattling off statistics with the best of them, but he can also wax so rhapsodic about dendritic architecture that he vaults into a sylvan version of surfer-speak: "Biodiversity takes on a whole new meaning, dude! It's mushrooms and salamanders and a new ecosystem up there. And gnarl! That gnarly oldness runs the show." Sillett subsequently built his academic career—he now holds a Ph.D. and is a professor of botany at Humboldt State University in Arcata, California— around his passion for redwood.

Jim Spickler is similarly fanatic. In 1995, at the age of twenty, Spickler was awarded a government contract to survey marbled murrelets living in the canopy of a redwood forest near Santa Cruz, California. He merged climbing techniques he had learned on cliffs in the Mojave Desert with those previously developed by tree researchers. "Climbing a redwood is more like climbing a mountain than climbing a tree anyway," he says. Like Sillett, Jim Spickler is an astounding athlete: he and I once climbed trees for nearly all of one tiring day, after which I headed for the nearest bottle of Corona, while Jim surfed the frigid waters of the midwinter North Pacific, then played soccer for three hours, until nearly midnight. When Spickler and Sillett met, it was a match of intellect, enthusiasm, and athletic talent made in heaven.

One rainy morning I drove north from Arcata to meet Steve and Jim at Iluvatar. Inside the car, the olfactory palette of civilized life surrounded me as I drove: the vinyl of the dashboard, my morning coffee and scones, the fruity scent of a freshly peeled banana, and the muddy nylon of clothing still wet from yesterday's work. In truth, the smells were so commonplace I barely noticed them. But when I stopped deep in the redwoods and pushed open the car door, a massive gust of evergreen fragrance—Christmas tree aroma multiplied a thousand times— overwhelmed me. I felt like a Magellan traveling around some Cape Horn of sensory experience, passing from an old, dull world into a glitteringly fresh new one. The vegetation, too, was beyond the commonplace. Knee-high sword ferns alternated with succulent expanses of cloverlike wood sorrel. Velvety moss a foot thick coated centuries' worth of fallen tree trunks, from which the champion trees of the future sprout. Then, of course, there were the immense trees vaulting into the mists. Enormous as they might look from the ground, they are, I learned through climbing, much taller than they appear.

Standing at Iluvatar's base and looking toward the heavens, I realized that the redwoods had been much more than a spatial or architectural revelation. We had ascended not just Iluvatar, but other trees, named Demeter, Zeus, Four Horsemen, Ballantine, Atlas, and Rhea. No longer were they nameless monuments of biology lost in an amorphous forest. They had become individual characters, each possessed of unique texture, color, shape, and personality. Just as with people, appreciating them had been a matter of time: the time to slow down, the time to look, the time to listen to what they said. It doesn't come quickly or easily, this awareness. One doesn't jump out of a car, snap a few photographs, then race off. Quiet, deliberate engagement is essential. The redwoods have been embedded in time for a long, long while, and their secrets take time to hear.

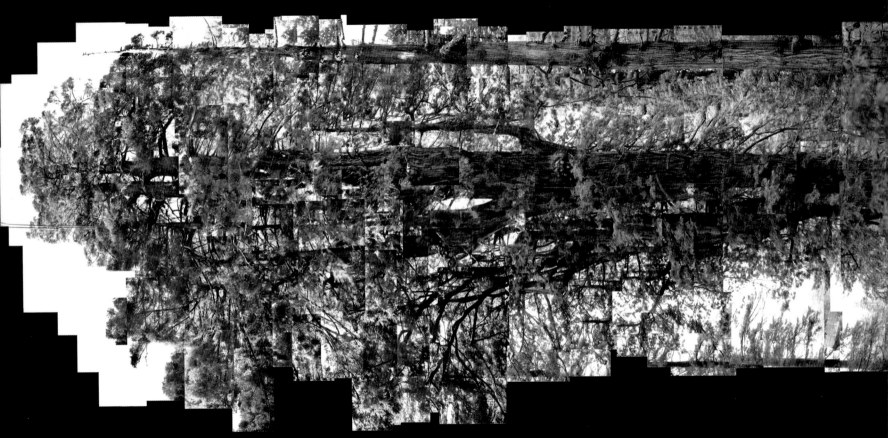

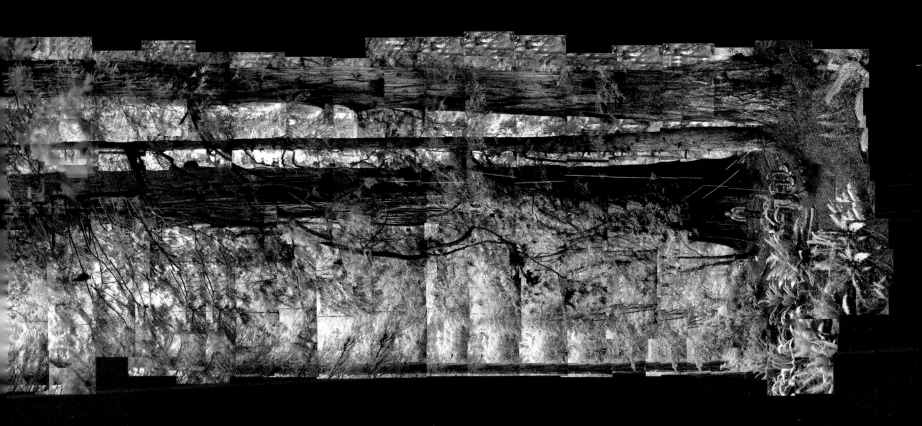

COAST REDWOOD
"Iluvatar"
Prairie Creek Redwoods
State Park, California

The Del Norte Titan, with 37,400 cubic feet of wood, was for several years believed to be the largest redwood in the world. Revised measurements early in 2004 put DNT, as it is known for short, somewhere in the pantheon of the largest four redwoods; ongoing scientific debate makes the exact ranking uncertain. Its exquisite bark makes DNT one of my favorite redwoods. But the best thing about the tree may be the story of one scientist's encounter with the invincible little mammal living in it.

Steve Sillett had climbed to the top of DNT, 307 feet above terra firma, to do his research. After he returned to the ground, he pulled down his 11-millimeter climbing rope. In its place he left a thin black parachute cord, which made one continuous loop from the ground on one side of the tree, up over the highest branch, and back to the ground again. The cord could be expected to stay in place for years. Most important, it was now his ticket back to the crown: by simply tying his climbing rope to one end of the cord and pulling on the other, he could quickly reinstall his climbing rope in the tree instead of going through the much more complex process of climbing the tree itself. Sillett went home content with a good day's work.

When he returned days later, the cord lay in a pile of spaghetti on the ground. He was stunned. With great effort, Sillett reclimbed the tree and placed another cord. The next time he came back, the wretched line was on the ground again. A Douglas squirrel or a marten, he now realized, must have been chewing through the line. Cutting it once was improbable, twice inconceivable. "I admit I had thoughts of killing whatever rodent was up there," this lover of all things arboreal confessed to me.

Up Sillett climbed again. This time, he looped a stainless steel cable—totally marauder-proof, of course—in place of the nylon cord. His system unfortunately had a tiny Achilles heel: one end of the cable didn't quite reach the ground, so he extended it with a bit of cord and anchored it to a sapling. No way, he thought to himself, will that animal come all the way down here to mess with it.

But the razor-toothed beast spoiled the best-laid schemes of man once more. The squirrel—or marten or whatever it was—hiked down the tree, gnawed through the short bit of chewable substance available, and yanked the cable out of the tree. With this, the scientist gave up. "I had hubris and thought I could beat him," Sillett says. "But he definitely won. He caused me so much physical and mental anguish, I was a wreck. He must have been some kind of guardian force for that tree. What's the point in going back up? I surrender."

Neither Sillett nor anyone else has ever climbed Del Norte Titan again. I did climb two adjacent trees, however. One gave me a viewpoint much like the squirrel's as he peered down from his castle ramparts at the alien invader.

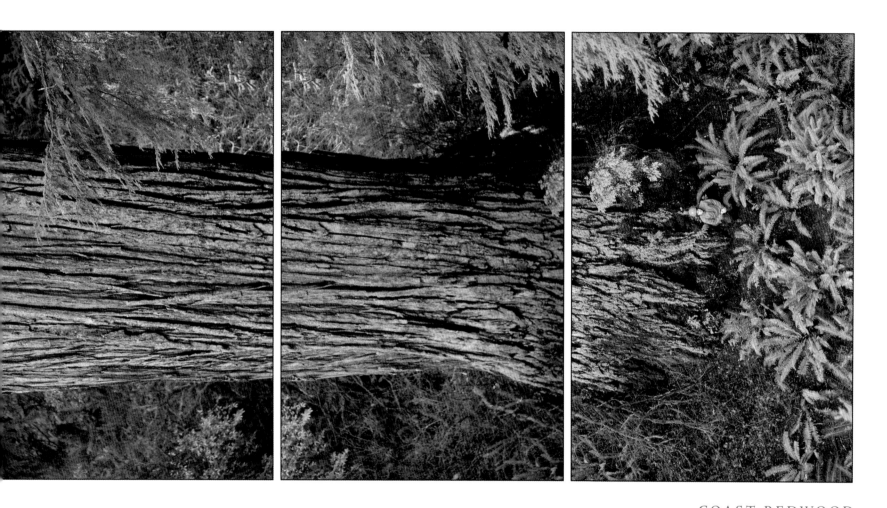

COAST REDWOOD
"Del Norte Titan"
Jedediah Smith Redwoods State Park, California

Stratosphere Giant is the Mount Everest of the plant kingdom. No living thing is taller than this tree. When I photographed it in September 2003, it was 369 feet 10^1/2 inches tall; as of 2005, it had grown 4 more inches.

Strat, as the tree is fondly nicknamed, lives in the Bull Creek Flats area of Humboldt Redwoods State Park, 20 miles inland from the Pacific Ocean in Humboldt County, California. Bull Creek holds the world's record for standing biomass. I had the privilege of seeing Strat only because Steve Sillett needed visual documentation of the tree and included me on his research team. He made me swear to keep its precise location secret, so admirers wouldn't come and love the tree—and the delicate habitat around it—to death.

No fewer than 115 other redwoods surpass 350 feet in height. Almost all of them grow on flat topography near streams, where moisture is abundant and the soils fertile and stable. Through time, the height of a given tree will fluctuate in response to changing climate; in drier periods, less water is drawn up the tree and the top starts to die back. Given the moisture, wind patterns, and other climatic variables prevailing in northern California during our time, redwood are apparently physiologically capable of growing up to 425 feet tall. Studies show that they can add 9 or 10 inches of height each year under ideal conditions.

The alluvial terrain so desirable to redwoods was, of course, also favored by loggers, who used the watercourses to float timber to sawmills. Over time, lumberjacks leveled something like 99.5 percent of the redwood in the valley bottoms. By all accounts, the trees that survive today are diminutive compared with those that existed a century ago. In the species' entire range, only 5 percent of the forest is virgin; the rest of the primordial forest is either regrowth or alder thicket. Redwood preservation started only at the turn of the twentieth century, half a century after logging had begun. A focus on identifying especially tall trees soon developed, but not until the 1980s did tree lovers Michael Taylor, Chris Atkins, and Ron Hildebrandt begin pinpointing many of today's stars. Atkins himself first found Strat. Sillett and Jim Spickler were the first to touch its highest fronds.

One drizzly morning, I hiked to the base of Strat with Jim Spickler, Steve Sillett, and Sillett's wife, Marie Antoine. A phenomenal pile of equipment erupted from our packs. We had well over a mile of cordage, everything from fishing line to a 13-millimeter rope with a breaking strength of 9,000 pounds. There were many dozen carabiners, 30 yards of nylon webbing, two laptop computers, apples, energy bars, turkey sandwiches, and a gallon of water. Athletic gyrations ensued. In a few hours, we had climbed to the top of Strat. On the way, Jim discovered the vacant nest of a marbled murrelet, a highly endangered black-and-white seabird with stubby wings, perched on the end of a small, bark-covered limb 320 feet above the ground (the highest murrelet nest ever found).

We rigged the same sort of Tyrolean traverse, complete with ascent rope tied to its middle, that we had devised for the Stagg sequoia shot. Then, using equipment cobbled together from big-wall rock climbing, caving, deep-ocean sailing, and tree science, I climbed upward. The higher I went, the more apparent it was that a cold front was blowing through. Treetops gyrated. I heaved up and down like a man in a rowboat at sea. By the time I topped out, I was suspended nearly 350 feet—thirty-five stories—off terra firma. I was 75 feet away from the nearest twig. I kept telling myself the Tyrolean could withstand phenomenal strain and I should be safe. But such windy conditions had always been my worst nightmare: the traverse was anchored to trees of different diameters, heights, and branch patterns, and neither of them flexed in the same way as they caught the wind. Perhaps even a falling branch, a flitting bird, or the stroke of a butterfly wing could slash the overstressed rope—or so I imagined in my darker moments. In any case, the rope was being subjected to terrible stress, with no safety backup if it broke.

Animal anxiety made me want to head down. But I was not oblivious to the beauty all around—and was very conscious of how much work had gone into getting me where I was. For ten bouncy, mildly seasick minutes I photographed Strat's crown, where Jim and Steve nestled in the branches. The worst gust of all howled through. Since my cavalier friends were lashed to a 6-foot-thick column of redwood, they could persuade themselves that the trees suspending my life in the air appeared to be in dynamic harmony and that I was safe. From my profoundly different perspective, dangling on what now seemed as flimsy as spider silk, our margin of error was way too thin. I set up a blue-and-red aluminum rappelling device called a Petzl Stop and zipped toward the ground as fast as I safely could. The metal of the descender quickly became too hot to touch. After a few minutes, my boot soles kissed sweet and deliciously horizontal earth.

Another sweat-soaked day passed as we set up a new traverse with a better perspective. The third day found us at Strat by dawn. We wanted to finish the photography before the wind picked up, the sun burned off the morning clouds, and the light became contrasty. I felt a calm resignation as I looked up the ascent rope: I had climbed mountains, rock walls, and trees for thirty-two years, and each adventure, I now realized, had been training for this ascent of ascents. No matter what ill winds might blow, I was where fate had meant to put me.

I headed for the sky. Not even the slightest zephyr swirled. Strat's tower of exuberant chlorophyll burst above the world; I was acutely aware of being in the presence of a ravenous life force. I photographed the crown, then began rappelling. Never before, I realized, had anyone seen what I was seeing.

The great tree was a blend of orderly form—the trunk—and dendritic chaos—the billion weavings of foliage, branches, and bark. According to classical artistic standards, the chaos should not have been attractive. By now, though, the trees had revealed that beauty transcends tradition and that the exquisite lives everywhere. Each frame of the forest seemed glorious. Eventually I slid the final 6 feet to the ground. Eight hundred fourteen frames (5 gigabytes' worth) had clicked through the viewfinder in five hours. During the entire descent, clouds had covered the sun and the light was ideal. Incredibly enough, the moment my feet landed on the root beer–colored carpet of needles on the forest floor, the sun gleamed through for the first time. Neither the scientists up in the canopy nor the photographer standing contentedly on the ground could miss the coincidence. Steve Sillett's voice crackled over our two-way radio: "Dude! The tree gods have been taking care of us."

Quercus chrysolepis
Springville, California

This national champion grows along Rancheria Creek, on private land in the western foothills of the Sierra Nevada. The trunk measures 345 inches in circumference, and the spread of its crown reaches 121 feet. Look for a child's swing in the bottom left corner of the picture to give a sense of scale.

The oak is at least several hundred years old. Members of the Yaudanchi tribe, Native Americans of the San Joaquin Valley who were herded onto a reservation in 1857, often camped under its sheltering limbs. A man named Mike now owns the land, and he found a spectacular array of tribal artifacts under the tree.

He spread them out for me on the tailgate of his Ford pickup. There were burnt-orange pottery shards the size of dinner plates as well as dozens of arrowheads of black obsidian and white quartz. He showed me hundreds of beads that the natives had acquired in trade from Spanish missionaries. Some were cobalt blue; white ones with red centers were called red hearts; handmade padre beads were shaped from red clay.

Mike also had an intact metate, a porous stone platter used to grind acorns and seeds. But his greatest find was an elegantly crafted buckskin quiver that was 18 inches long and held six arrows with tiny obsidian points as sharp as razors. An ultrafine twine bound little gray feathers to the wooden arrow shafts, all of it cemented in place with an amber resin still rich with pine scent. The quiver and arrows, Mike believed, belonged to a child. Presumably, the child's family meant to revisit the campsite, but fate had other plans.

COAST REDWOOD
"Stratosphere Giant"
Humboldt Redwoods State Park, California

HONEY MESQUITE

Prosopis glandulosa var. glandulosa

Leakey, Texas

Mesquite is one of the dominant arboreal species along the southern edge of the United States from Texas to Arizona. In most places, it looks more like a bush than a tree. But this mesquite, growing alone in an open field, breaks the mold at 55 feet in height with a crown spread of 89 feet.

My original plan for supporting the 80- by 100-foot background failed. Just when all hope was lost, three locals—Monty Pannell, his son Jeff, and their buddy Wayne Rushing—became captivated by what I was doing and donated two days helping me get the shot. "The picture seems like a right nice idea," one of them laconically drawled. These good ol' boys not only had access to the kind of crane called a cherry picker, they possessed the skills necessary for welding together an 80-foot beam. We lashed the background to the beam and the beam to the cherry picker, and winched the whole contraption into the air. In the slightest zephyr, the fabric billowed like the sail on a square-rigged clipper ship and nearly pulled the crane over. Monty shows up as a shadow on the bottom left of the frame.

PLAINS COTTONWOOD

Populus deltoides var. occidentalis

Hygiene, Colorado

Cottonwood usually survive no longer than a century, yet the species lives large and fast. The trunk of this national champion, for one, is so big that a decent-sized car could fit inside. It grows in a wild bottomland at the eastern foot of the Colorado Rockies, a half-hour drive from my house. I tried many times to photograph this tree in a snowstorm, but every time big storms hit I was far away chasing some other chimera. The few storms that arrived when I was at home dumped feet of snow on our house but nothing on this tree 20 miles away.

I have learned to live with, if not embrace, such vagaries of the creative process. Conscious desire yields no more than a starting point of an image; in the end, only the pictures that are meant to be come into existence. How fate decides such things is impossible to know: at a minimum, one's life seems to have a finite number of intersections with a given subject, so going back to the same subject over and over again hoping for slight improvements tends to do little good.

All of which is a way of pointing out that I made this image on my only springtime visit to the cottonwood. Obsessed with my preconception about photographing the tree in winter, I didn't notice this powerful diptych on the contact sheets at first. Only after I looked at the film with a fresh eye did I realize the tree had already given me the image I was meant to have.

LIVE OAK
"Angel Oak"
Johns Island, South Carolina

LIVE OAK

Quercus virginiana var. *virginiana*

"Angel Oak"

Johns Island, South Carolina

The Angel Oak is named not for heavenly hosts but for a family who owned this section of the Carolina coastal plain early in the nineteenth century. Ancient live oak usually have hollow trunks, making precise dating of this tree futile. Educated guesses put its age at around 1,400 years. If so, it germinated when King Arthur and Muhammad were afoot. One of the oak's enormous limbs sprawls 90 feet from the trunk.

Though my affinity for trees is obvious, I don't usually indulge in romantic, druidic speculation about them. When I'm in the presence of trees like this oak, though, I can't help wonder who truly is the observer and who is the observed. While we watch them, do they gaze back at us? It is easy to understand the reciprocity of the visual exchange when a creature has eyes. Does a plant have some other sense that we mammals lack the capacity to understand?

This sycamore is the largest tree in New England, with a girth of more than 25 feet and a crown spread of 140 feet. It was named after forester Gifford Pinchot. Born into a wealthy Connecticut family and Yale-educated, Pinchot studied forestry techniques in France, where he became convinced that forests should be logged selectively, not clear-cut. The notion was a radical one at the time, and, at least in American forestry circles, it still is.

In 1898, Pinchot took the reins of a government bureau soon to be named the U.S. Forest Service. He and President Teddy Roosevelt, a longtime friend, did not believe in allowing private interests to run amok in their use of wild lands. In March 1907, Congress enacted legislation limiting Roosevelt's ability to protect any more territory. The night before the bill became law, Roosevelt and Pinchot, along with Pinchot's assistant Arthur Ringland, sat down with maps of the western United States. Using a blue pencil, they marked off 16 million acres of the West and, by executive fiat, put it under the jurisdiction of the Forest Service.

Today, the Forest Service is seen by many as an agent of land abuse more than one of conservation. The debates it has been involved in over the past century will likely go on for many, many more. In any case, without the vision of Pinchot, his two friends, and that blue pencil, America the beautiful would today be much less beautiful than it is. They understood that a sane society couldn't forget aesthetics even as it focused on the utilitarian.

SYCAMORE

"Gifford Pinchot Sycamore"

Simsbury, Connecticut

LIVE OAK
"Middleton Oak"
Charleston, South Carolina

LIVE OAK
Quercus virginiana var. *virginiana*
"Middleton Oak"
Charleston, South Carolina

Known as the Middleton Oak, this tree has sprouted toward the sun for more than a thousand years. One of the oldest and largest living organisms in the eastern United States, the tree grows on a preserved eighteenth-century plantation called Middleton Place, which once grew cotton and indigo, a shrubby legume used to dye clothing deep blue, for the textile trade. The estate was the home of Henry Middleton, president of the First Continental Congress; his son Arthur, who signed the Declaration of Independence; and Henry's great-grandson, Williams, who signed the Articles of Secession that began the Civil War.

The oak grows along the banks of the Ashley River. The moments that must have unfolded as people and goods were once bought and sold along this waterway can only be imagined. I was moved by the elegance of that riverine canvas behind the reaching arm of the oak.

LIVE OAK
"Middleton Oak"
Charleston, South Carolina

PLAINS COTTONWOOD

Populus deltoides var. *occidentalis*

Boulder, Colorado

Though not a particularly gigantic specimen (its base is 32 inches in diameter), this cottonwood grows in my backyard. I lifted my eyes to it every few minutes as I wrote this book, in fact. The tree was planted four decades ago, when our tiny piece of the Great Plains had completed a 120-year process of being converted from thundering buffalo herds to pasture to home site. Each year, the tree metamorphoses from winter dormancy to springtime optimism, and each year the process seems miraculous.

One year, I finally decided to put the process on film. I drove a small crane into our yard and winched a billowing fabric background up behind the tree (perhaps gaining a reputation as the neighborhood crackpot). The two pairs of images were exposed from an upstairs window of our house.

LIGNUMVITAE

Guaiacum sanctum

Key West, Florida

Lignumvitae produces incredibly dense, heavy wood, almost like iron. A cubic foot weighs 88 pounds (ebony, another famously dense wood, weighs 73 pounds per cubic foot; mahogany weighs in at 45). Lignumvitae is so strong and durable that it is fabricated into moving parts for ships and industrial mills.

According to one published source, this national champion lignumvitae may be 1,600 years old. But since trees in the year-round growing season of the tropics lay down no annual rings, its true age is unknowable. In any case, this lignumvitae appears to be one of the few pre-Columbian trees in the Florida Keys; virtually all the old-growth, including once-towering stands of mahogany, was cut by Spanish colonists and those who followed. The tree stands in the courtyard of St. Mary's Roman Catholic Church; the yellow buildings are townhouses on the other side of the churchyard's perimeter wall. Hurricane George, which clocked 105-mile-an-hour winds, blew off the lignumvitae's crown in 1998.

As we set up this shot, a cold front surged through the Keys. A frantic wrestling match ensued between my desire for a picture with a fabric background and the wind's desire to carry away all movable objects. There was an ever-present danger of breaking a window, shredding the background against the aluminum gutters of the building, and even pulling over the crane that held up the right side of the background. The dynamism of the moment somehow infused the image with a haunting, surreal quality.

FLORIDA ROYALPALM

Roystonea elata

Fakahatchee Strand State Preserve, Florida

The Fakahatchee Strand of southwestern Florida is hardly a quotidian landscape. Viny thickets, rattlesnake-infested sawgrass meadows, and swamps slithering with cottonmouths abound. Florida panthers hid in the preserve's 63,000 acres before the purebred species went extinct. When armadillos escaped from a leprosy experiment at a Jacksonville hospital (400 miles away), their descendants migrated to the Fakahatchee like religious pilgrims seeking the promised land; their breeding success shows they found what they were looking for. And one patch of forest supports three thousand royalpalms, the largest concentration of the species known in the world.

Taxonomically, most palms are considered grass, not trees. Regardless of this technical distinction, Dan Ward, dean of Florida botany, includes royalpalm in his encyclopedic study of the state's big trees because they are such a prominent, treelike feature of the regional landscape. This cochampion was surrounded by dense forest. A wider perspective was impossible, so I let the image be about the ground-level thicket as much as it is about the palm.

JUMPING CHOLLA

Opuntia fulgida

Pinal County, Arizona

One of Arizona's big-tree fanciers, a shy, quiet man named Mike Hallen, met us in the parking lot of a community college south of Phoenix. Mike had found a new champion cholla just a few weeks before my arrival. To find it again, we cruised for an hour and a half down rural back roads. Cotton fields alternated with wasted-looking earth where farmers overused and then abandoned the soil. Only bleak saltbush managed to hang on here and there. We passed tiny settlements, each just a few single- or double-wide prefab houses; the colors were limited to gray, tan, or peach. Junked cars and tumbleweed abounded.

We continued driving. Living desert finally returned: shades of green and yellow and russet and ivory. We turned off the blacktop onto a road of tangerine sand. Cholla sprouted in such profusion that the place seemed like the desert equivalent of a jungle. Mike's newly found champion was embedded in a landscape of so many plants that I wondered how he found it at all.

The rigid spines of cholla are hair-thin, yet astoundingly sharp. Even when you think you're nowhere near the plant, spiny clusters still manage to spear anything within reach, thus the moniker "jumping." The fiendish things are capable of penetrating car tires, the hard insulation on electric wires—and a knuckle on my left hand. To accent the champion cholla in the midst of its surroundings, I decided to use a muslin backdrop.

JOSHUA-TREE

Yucca brevifolia

Mojave, California

This species was named by Mormons migrating across the desert from San Bernardino, California, to Utah. The humanoid tree shapes evoked for them the biblical seer Joshua pointing to the promised land.

Joshua-trees naturally live in the parched expanses of the Mojave Desert, where average annual precipitation is less than 5 inches. They sometimes don't feel a raindrop for years. But this, the largest Joshua-tree in the United States (height 41 feet, trunk circumference 147 inches), inhabits a different planet: a community golf course where the grass is injected with so much chemistry that it practically glows. The golf course, tree included, soaks up 11 million gallons of water each month from the beginning of July to the end of September; in a whole year, it consumes 60 million gallons.

Though the Joshua-tree is sculpturally attractive, as so many of its kin are, I found myself less than elated about a tree bulked up on the botanical equivalent of steroids. Still, its story is significant, so I include it in this portfolio.

Odd as it might seem, saguaro are considered trees, since an internal wood skeleton holds up the pulpy, spiny mass of the plant. The biggest can grow 50 feet tall, store as much as 9 tons of water, and live 250 years. No fewer than six saguaros, scattered across a 150-mile stretch of Sonoran desert, were considered national cochampions on the day we went looking for them.

One of the champions grew in remote desert two hours northwest of Phoenix. Our truck bounced over chassis-scraping granite knobs, through billowing dust, and around washboard turns. We were grinding slowly through a tan canyon when our guide, Joe Plaggenkuhle, told us to stop. Somewhere on the slope to our right, he said, stood a saguaro named Condor. We crossed a sand wash, scrambled up the hill, wove through olive-colored mesquite, and kept on the alert for rattlesnakes.

Plaggenkuhle wondered aloud why the place didn't look right. After poking around for a few minutes, he realized the answer: Condor lay on the ground in shriveled brown and gray chunks. A downdraft from a summer thunderstorm had presumably blown it down. Disappointed, we headed back to town.

That evening, Bob Zahner, the dean of Arizona big-tree hunters and a retired professor of forest ecology, met us at a Mexican restaurant so tasty that I wolfed down two servings of enchiladas. Zahner drew a map to his favorite saguaro, in a suburb northeast of Tucson. The next day, we found the two-century-old cochampion sandwiched between a busy street and a coral-colored enclave of duplexes.

Through a Florida horticulturist, I learned of this tree hidden in a backyard a few blocks from the Atlantic Ocean. Mangrove are generally found in aquatic habitats, so this tree likely didn't grow there naturally, but was instead planted by a homeowner long ago.

The house behind the tree was distracting, so I draped it with a huge piece of background fabric. Wrangling that sort of prop, setting up lighting, and photographing such a large subject normally makes for a long workday, with many experiments needed to get the image right. But our labor went uncommonly fast, requiring only a few minutes of test shots before I exposed the final frames. It seemed as if the image already existed and was just waiting for us to come along and record it.

Bristlecone pine are the oldest living organisms known. The eldest, a tree named Methuselah in honor of the biblical patriarch who is said to have lived 969 years, is officially 4,734 years old. Certain subtleties of ring-counting procedure may mean it is actually more than 5,000 years old. This book shows Methuselah and other 4,000-plus-year-old bristlecone, but to protect Methuselah from souvenir hunters, I will not identify the trees specifically.

Tree longevity is often compared to human history. Methuselah sprouted around the time the great Egyptian pyramids were built. I also did some anthropocentric arithmetic and arrived at a startling realization: Methuselah germinated at the time of progenitors roughly 250 generations older than I am. The tree is so old that a 2-foot-thick layer of the mountain beneath it washed away during its life, and roots now perch in empty air as a result.

Bristlecone are scattered through western mountain ranges at an altitude of roughly 10,000 feet. Methuselah and most of the other superannuated individuals live in the White Mountains of east central California. Growing conditions would seem to be atrocious. Water is minimal to nonexistent, averaging 10 inches per year. Soils are stony. Wind and blowing snow lash the trees. During drought years they hardly grow at all, and their annual growth rings are no thicker than a human hair. Bristlecone live frugally, keeping alive only a minimum of branches and bark. Their trunks become weather-beaten into fantastic shapes and colors, and their wood is extremely dense, full of resin, and resistant to insects and disease. As bristlecone scientists are fond of saying, adversity breeds longevity.

Dendrochronologist Edmund Schulman discovered Methuselah in 1957. According to one of Schulman's field assistants, Doug Powell, Schulman had an ailing heart and "hoped some substance could be distilled from these old trees that a human being could somehow absorb and [use] as a factor in human longevity." Unfortunately, Schulman's heart gave out when he was forty-nine, a year after he discovered Methuselah.

Another bristlecone was named after the Greek god Prometheus, who is said to have been shackled to a mountain and repeatedly gnawed by an eagle. In 1964, graduate student Donald Currey was studying ancient glaciation cycles on Nevada's Wheeler Peak by surveying bristlecone. He and a friend chainsawed down "the first old tree we climbed to on the crest of the lateral moraine. [Only] five minutes of looking was involved." It was Prometheus. They sliced off a cross-section and hauled it back to a motel room to count its rings.

Nearly four decades later, on a quiet night in Bishop, California, with the Milky Way blowing clouds of stars across the desert sky, I followed a retired forest geneticist named LeRoy Johnson into his tidy garage. He lifted a gray metal box off a shelf. Inside this reliquary was a piece of orange-tan wood 2 feet long, 1 foot wide, 8 inches thick, and sanded smooth as silk. LeRoy offered it to me sacramentally, reverently, in the palms of his upturned hands. It was one of the pieces of Prometheus on which Currey had counted 4,844 growth rings. In the name of science, Currey had killed the oldest single living organism ever known.

Above and opposite:
INTERMOUNTAIN BRISTLECONE PINE
White Mountains, California

Right and opposite:
INTERMOUNTAIN
BRISTLECONE PINE
White Mountains, California

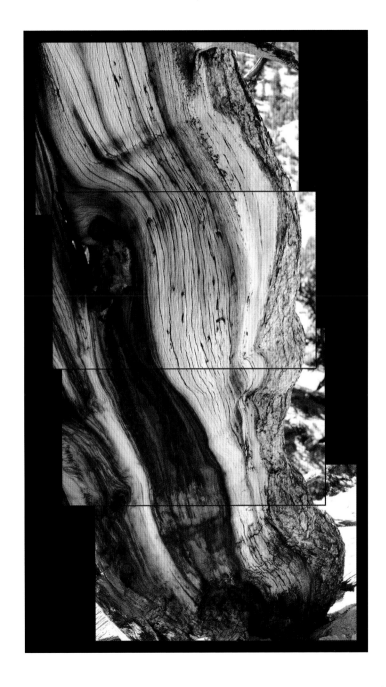

The secret life of distinguished trees is in a far greater state of flux than I had ever imagined. During the years I worked on this project, a surprising number met their end. The champion staghorn sumac, which grew at a school in Alabama, was cut because it had become diseased. During a winter gale, a titanic redwood I had counted on to anchor the ropes for my shot of Stratosphere Giant crashed to earth. The champion white oak, saguaro, and chestnut blew down in storms. A lightning strike, a tornado, the ignorance of a highway maintenance crew, and a vandal have finished off other champions in recent years.

The basswood shown here had been a 116-foot-tall national champion. Mike Davie, one of the Southeast's leading big-tree aficionados, took me to find it in a remote hollow of the Smoky Mountains, having no reason to believe it was anything but alive and well. But when we got there, we found it toppled.

A tree falling in the forest offers the classic existential enigma: if a tree falls in the forest and there's no one there to hear it, does it make a sound? Since none of us ever seems to hear a big tree falling, I'll pass on the remarkable experience of my friend Marie Antoine. Marie and some friends were sleeping in hammocks high up in a northern California redwood. The night was perfectly windless, so there seemed no reason for a tree to fall. But suddenly, mayhem broke out nearby.

"It sounded like gunshots as the trunk of a tree snapped," she told me. "Then there was a loud rumbling. You couldn't tell where it was coming from except that it was getting closer. As it started to fall through the crowns of other trees, there was an unimaginably loud crashing and cracking—unbelievably loud. Then it hit the ground—a big *whoompf*—and the earth literally shook. Our tree shook, too. It was like an earthquake. Then it was silent. We went down the next morning to look for it. It turned out to be a Sitka spruce, about 300 feet tall, that had fallen 1/4 mile away. Who knows why it fell? But the entire crown buried itself in the mud. The rest of the tree literally exploded into bits."

162

Bark is a waterproof skin of dead cells that protects a tree's interior plumbing. The furrows in bark are essentially stretch marks that appear as the tree expands. On some trees, like giant sequoia, bark can be 12 or more inches thick—good insulation from wildfire. Other species, like birch and aspen, have skins just $^1/_4$ inch thick. In the famously exquisite bark of paper birch, cells are laminated on top of one another in layers, so the bark can stretch without furrowing. Microscopic air spaces riddle the outer layers, reflecting light in all directions and making the bark appear white. Should the inner, reddish brown bark be exposed, a black scar develops.

Rain blasted in cold sheets off Lake Michigan as we made these exposures. It was all I could do to find a satisfactory camera angle, keep my strobes functioning, and expose a few frames. Not until weeks later did someone notice a little polar bear face peering out from the tree.

MOUNTAIN PAPER BIRCH
Leelanau, Michigan

QUAKING ASPEN

"Pando"

Fish Lake, Utah

QUAKING ASPEN

Populus tremuloides

"Pando"

Fish Lake, Utah

This aspen grove consists of forty-seven thousand genetically identical trees, each sprouting from the same root system and each a clone of its neighbor. The grove is named Pando, Latin for "I spread." Collectively, the trunks are the heaviest single organism known in the world today.

Pando thrives in a small, relatively moist upland in central Utah. It covers 106 acres and weighs at least 13 million pounds. (The General Sherman giant sequoia, the most massive single tree in the world, weighs about a third as much.) Enormous aspen groves like Pando, one botanist has proposed, may be self-perpetuating and reach an age of a million years or more.

Photographing a single subject with forty-seven thousand parts was no small challenge, especially because I wasn't interested in making a traditional decorative aspen picture. I visited Pando in the fall and in midwinter. Eventually, I hit upon the idea of moving through the forest while looking simultaneously in different directions.

The eighteen-column picture on pages 168–169 shows the camera looking both forward and backward as I walk through Pando. I picked my traverse in response to the sunlight and shadows. Partway through the process, it became clear that fate was guiding me through a robust little juniper and toward a decrepit old aspen. Just as I reached the aspen, the memory in my digital camera ran out. I had previously been brooding about aging and mortality (it was the year of my fiftieth birthday), so this image seemed like a performance piece about the arc of human life.

Aspen are not completely sheathed in the layer of dead cells known as bark. Their smooth trunks are actually living tissue capable of photosynthesis. This living parchment is irresistible to graffitists, apparently because trees are perceived as outlasting the carver. (Incidentally, I know two particular aspen in the Colorado Rockies that black bears love marking with their claws.) Inscribing the heartthrobs of adolescent love on trees has an especially long tradition: it presumably signifies possession of the beloved and symbolizes hope that love will vicariously acquire the endurance of sylvan life. I am not advocating carving tree bark. But it seemed fitting to include a sample of the Pando inscriptions as a study in how humans relate to nature.

SUGAR MAPLE

Acer saccharum

Lenox, Massachusetts

Big-tree hunters find many champions in places like cemeteries, churchyards, and school campuses. Trees in institutional landscapes benefit from pruning, watering, and artificial feeding—and, most important, are protected from chainsaws. This spectacular maple (not a national champion, but exceptional for its region) grows next to a building called the Church on the Hill, a United Church of Christ congregation first assembled in 1769. Tombstones checker the earth on the west side of the tree.

The maple was flush with autumn gold when I had the delectable pleasure of rappelling through its foliage. Several pigments are dissolved in the sap of deciduous trees. When the cold of autumn comes and the abscission cells at the base of the leaf stem block sap flow from tree to leaf, the green of chlorophyll gets shut off first. What happens next depends on the species and chemistry of an individual tree. A pigment called xanthophyll might predominate, coloring a leaf yellow. In other cases, carotene will dye a leaf orange and red, or anthocyanin will turn it purple. The random alchemy between all three pigments creates dazzling rainbows of variegated color.

As I descended, my camera was within inches of the maple's trunk and the xanthophyll-dyed leaves. At such close proximity, it would have taken thousands of frames to concoct a continuous mosaic. Instead, I sampled elements of the scene and gave them coherence by using a grid pattern.

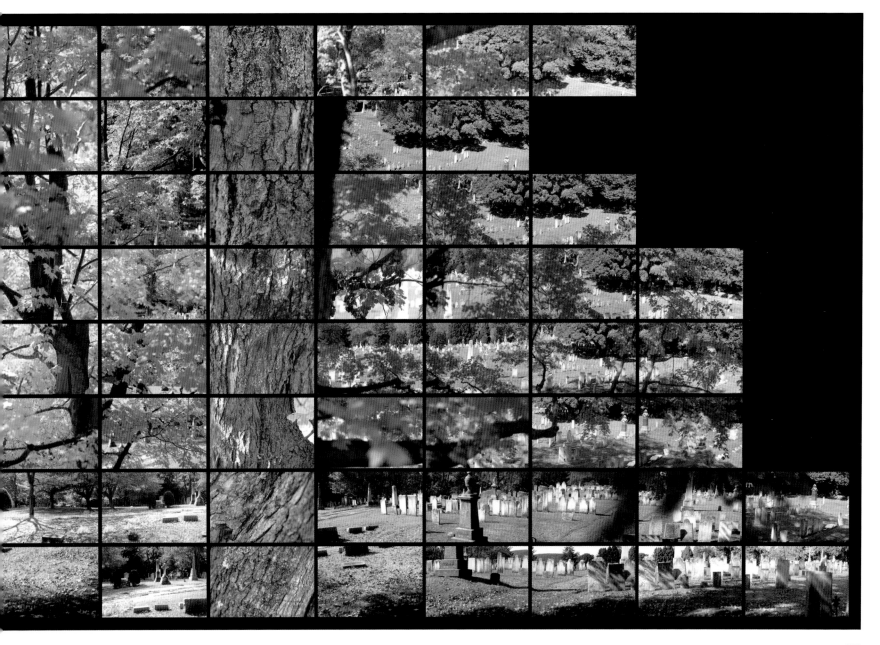

ENGELMANN SPRUCE

San Juan National Forest, Colorado

ENGELMANN SPRUCE

Picea engelmannii

San Juan National Forest, Colorado

The tree farthest right in the panorama on pages 174–175 is the largest Engelmann spruce known in Colorado. It lives at 10,640 feet. A storm swirling in from the Pacific had just swaddled the Rockies in 15 inches of powder. The wintry landscape tempts me to report an intriguing connection among trees, music, and photography.

Classical-music lovers have long marveled at the rich sound of Stradivarius violins. The Italian master used spruce from the Alps to make his instruments. It has recently been discovered that these spruce lived between 1625 and 1720, the coldest part of the Little Ice Age (a period of climatic cooling from the mid-fifteenth century to the mid-nineteenth century). Cool temperatures led to slow growth and exceptionally dense wood. And it was this dense wood that endowed the violins with their heavenly resonance.

These Engelmann spruce, living in conditions much like those of the violin trees, probably have dense wood and could make beautiful music, too—though I trust no one will ever be allowed to find out. In any case, a violin is to a musician as a camera is to a photographer: one tries to acquire the best instrument possible, practice with it for so long that it becomes second nature—and then play without thinking about it at all. Aesthetics, not technology, is the essence of art.

NORTHERN TREELINE
Hudson Bay, Canada

Treeline is not truly a line. Instead, it is a zone where the daily struggle against subzero cold, howling wind, and the sandpaper effect of blowing snow makes trees progressively smaller and more scattered, until eventually none can survive at all. The height or latitude of treeline fluctuates depending on regional weather conditions. Treeline is at 11,500 feet in the Colorado Rockies but only 4,500 feet in New Hampshire's White Mountains. Similarly, while trees fade out along the western shore of Canada's Hudson Bay at roughly latitude 58 degrees north, they manage to hang on to 69 degrees in northwestern Canada— nearly all the way to the Arctic Ocean.

One November, I was filming polar bears along the western shore of Hudson Bay, near Churchill, Manitoba, when these sturdy little spruce caught my attention. I drove a long way to look for more northerly trees but never found any. I include them as representatives of the millions of trees at arctic and alpine treelines, surviving with superlative endurance in the face of adversity.

BIBLIOGRAPHY

An enormous number of sources, both published and unpublished, helped guide the selection of trees for this book. One of the most important was the National Register of Big Trees, compiled by American Forests (www.americanforests.org). The definitive source on the conifers of the Pacific Northwest is Robert Van Pelt's *Forest Giants of the Pacific Coast* (Seattle: University of Washington Press, 2001).

For more location details, as well as perspective on botanical details and historical ecology, see the following books:

Cronon, William. *Changes in the Land: Indians, Colonists, and the Ecology of New England.* New York: Hill and Wang, 1983.

Dietrich, William. *The Final Forest.* New York: Penguin, 1992.

Irland, Lloyd. *The Northeast's Changing Forest.* Petersham, Massachusetts: Harvard Forest, 1999.

Kershner, Bruce, and Robert Leverett. *Sierra Club Guide to the Ancient Forests of the Northeast.* San Francisco: Sierra Club Books, 2004.

Noss, Reed, ed. *The Redwood Forest: History, Ecology, and Conservation of the Coast Redwoods.* Washington, D.C.: Island Press, 2000.

Platt, Rutherford. *1001 Questions Answered About Trees.* New York: Dover Publications, 1959.

Rajala, Richard. *Clearcutting the Pacific Rain Forest.* Vancouver: UBC Press, 1998.

Thomas, Peter. *Trees: Their Natural History.* Cambridge, United Kingdom: Cambridge University Press, 2000.

Van Pelt, Robert. *Champion Trees of Washington State.* Seattle: University of Washington Press, 2003.

Ward, Dan, and Robert Ing. *Big Trees: The Florida Register.* Florida Native Plant Society, 1997.

Williams, Michael. *Americans and Their Forests: A Historical Geography.* Cambridge, United Kingdom: Cambridge University Press, 1989.

A privately published monograph, *California's Biggest Trees*, is available directly from its author, Art Cowley (1681 West Memory Lane, Porterville, California 93257. Telephone: 559-784-3408).

Another excellent source of information is the Eastern Native Tree Society (www.uark.edu.misc/ents).

INDEX TO TREES

ACKNOWLEDGMENTS

By the time this project was completed, the energies of an extraordinary number of good-hearted people had been invested in it. To everyone I am hugely thankful.

Steve Sillett opened the incredible trove of the redwood forests. He and Jim Spickler were terrific allies during many wonderful days in the canopy. Their skill, enthusiasm, and sweat were indispensable; they also taught me tree-climbing techniques I used elsewhere in the country. Billy Ellyson was a crucial partner during certain epic days in both redwood and sequoia. Steve's wife, Marie Antoine, Nolan Bowman, and Clint Jones also helped move me up the magic beanstalk.

Washington, D.C.-based American Forests has kept a list of the biggest trees in America since 1940. This registry was the original framework for identifying my subjects. Executive director Deborah Gangloff and the rest of her staff were essential in pointing me toward the right trees—and the people who could find them.

The knowledge stored in the minds of a few incredibly passionate people should rightly be the subject of many books beyond this one. Forestry researcher Bob Van Pelt seems to know everything worth knowing about trees in the Pacific Northwest, not to mention a considerable fraction of the rest of the world. Bob Leverett, a leading light of the Eastern Native Tree Society, was tremendously helpful in revealing the secrets of New England. For that matter, different regional specialists handed me the keys to their kingdoms: Art Cowley in California; Bob Zahner in Arizona; David Milarch and Woody Ehrle in the upper Midwest; Leroy Johnson in bristlecone territory; Dan Ward and Terry Mock in Florida; and Mike Davie in the Great Smoky Mountains.

I'm not sure he would be comfortable being mentioned in the company of so many tree huggers, but retired logger Sonny Rouch occupies a special place in this series: as the patriarch of the family that owns the only big sequoia surviving on private land today, Sonny made it possible for me to photograph Stagg. Still others owned great trees or provided guidance in finding them: Mike Cobb, Vicky and Larry Dubuke, Charles Duell of Middleton Place, Bobby and Sheila Fetzer, Mike Hallen, Jerry and Nancy Holmes of the Circle Z Ranch, Jeff and Monty Pannell, Joe Plaggenkuhle, Wayne Rushing, LeAnn and Anthony Sharp, and St. Mary Star of the Sea Catholic Church in Key West, Florida.

Having an interesting concept is one thing, funding it another. If I lived in an ideal society—say, one where less collective wealth drained into vacuous entertainment or military hardware and instead went into the arts—I would have photographed more trees than I did. Most of the project was self-supported, so I am especially grateful to the partners who had the vision to come along for the ride. The first were *Vanity Fair's* Graydon Carter, Susan White, and David Friend (more on David later). Without the magazine's seed money, the project would never have germinated. The incredibly talented art department of *Audubon*, Kevin Fischer and Kim Hubbard, ensured that editor David Seideman would commission work in the southeastern United States. Barry Tannenbaum of *Nikon World* published an early glimpse of the series. Steve Petranek and Maisie Todd of *Discover* sent me to one last species, the chestnut, before the clock ran out and the book went to press.

John Rasmus, editor in chief of *National Geographic Adventure*, and his terrific staff, particularly Jim Meigs, Julie Curtis, Kalee Thompson, and Sabine Meyer, were captivated by the yarn about the making of Stagg's portrait and gave it wonderful space in their magazine. David Schonauer presented the trees elegantly in *American Photo* (he also superbly edited an early draft of the introduction). Kirk Brown, art patron extraordinaire, put wind in my sails at a crucial time. The staff at Leopard Communications, particularly Sherri Leopard, moved things along, too. Janette Gitler and Lew Colemen were essential in helping put the canyon live oak on film.

My darkroom partners, Photocraft Laboratories, worked hand in glove with me on many phases of post-production. I am especially grateful for owner Roy McCutchen's steadfast enthusiasm, Mike Harrison's quick fingers in Photoshop, and John Botkin's clear eye for color. Nikon and its staff, particularly Richard LoPinto, Bill Pekala, and Steve Heiner, gave considerable technical support. In the field, strong backs and good cheer came from Aaron Hoffman, Peter Holcombe, Stephen Miller, John Lichtwardt, Matt O'Keefe, and Jared Milarch. John Weller, Mark Johnson, and Kevin and Kelly Anderson helped with important technical support back at the studio.

Moral support was much needed and appreciated. Photographers Jim Nachtwey and Chris Anderson, museum curator Anne Tucker, artist Chuck Forsman, photo agent Bob Pledge, éminence grise Rich Clarkson, and my old climbing partner Jon Waterman each said the right things at key times. Chuck's input was essential one pivotal Sunday night as we finished the picture edit. Jose Azel, Dave Barry, Jane Fudge, Scott Roche, Jeffrey Smith, Dave Tippits, and Lauren Wendle helped keep my psyche afloat in their own different ways.

The graphic template was originally produced by Bob Morehouse and his terrific staff at Vermilion Design in Boulder, Colorado. Even with their beautiful maquette, publishing the book was difficult: historical trends, the choppy economy, and a society focused on armed conflict worked against it. For depressingly long periods, I wondered if a book would ever see the light of day. But then the angel of Barnes & Noble descended from heaven. The commitment at all levels of the organization, starting with CEO Steve Riggio, publishing president AlanKahn, and trade book division publisher Michael Fragnito, was exceptional. Creative director Jeff Batzli wove a graphic flow through diverse photographic styles and guided designers Astrid Lewis Reedy and Wendy Ralphs to our final destination. Editor Susan Lauzau kept the project on track and my writing in check. Richela Morgan and Karen Matsu Greenberg ensured the printing quality.

I save a few people for special mention until last. My parents, Jim and Alvina, steadfastly lent their support—in spite of not always understanding just why a project like this needed to happen. John Wiltse was a strong and steady presence through much of the work, a field assistant and friend par excellence. Similarly, the project would have been impossible without the sisterly devotions of my longtime office assistant, who goes by the name of Sport. Had she not organized the files, guided the traffic of images, and lent her touch to a host of graphic design issues, especially in the last hectic months leading up to press time, the book could not have happened. And without David Friend's faith, the project would have died a thousand deaths. By generating the initial funding, he helped turn an idea into reality. Later, during my darkest, most doubtful hours, he spurred me on, believing this project had to be completed no matter what the odds against it seemed to be. He became a brother and a comrade.

Finally, there is my immediate family. My wife, Suzanne, and my daughters, Simone and Emily, endured tremendous strain as *Tree* came to life. From the absences while I was in the field to financial difficulties to enduring my dour moods as manifold problems arose, the women in my life paid a real cost for these images. To all three I am humbly, eternally grateful. And to them I dedicate this book.

One last thought: as I've said elsewhere in the text, the trees were an eerily powerful presence throughout the project. Without their accord, much less would have come from my encounters with the trees. For their acquiescence, for their dignity, and for giving me safe passage, I thank them.

181

Some of the activities described in this book are inherently dangerous, and were performed by experienced climbers who had the requisite skill, training, and equipment to climb trees safely. We don't advise or recommend that readers attempt any of the activities depicted. The author and publisher specifically disclaim any liability resulting from the use of any description or information in this book.

Visit www.jamesbalog.com for other work by James Balog.

A contribution has been made in support of the effort to conserve and replant American forests.

Published by Sterling Publishing Co., Inc.
387 Park Avenue South, New York, NY 10016
©2004 by James Balog
Distributed in Canada by Sterling Publishing
c/o Canadian Manda Group, 165 Dufferin Street
Toronto, Ontario M6K 3H6
Distributed in Great Britain by Chrysalis Books
64 Brewery Road, London N79NT, England
Distributed in Australia by Capricorn Link (Australia) Pty. Ltd.
P.O. Box 704, Windsor, NSW 2756, Australia

ISBN 1-4027-2818-2

Color separations by Bright Arts Graphics Singapore
Printed and bound in China by Asia Pacific Offset
All rights reserved

1 3 5 7 9 10 8 6 4 2